The Story of Vermont

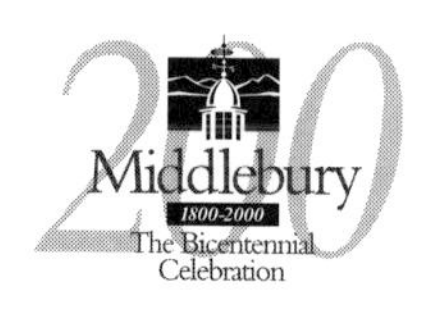

MIDDLEBURY
BICENTENNIAL
SERIES IN
ENVIRONMENTAL
STUDIES

THE STORY OF VERMONT

A Natural and Cultural History

Christopher McGrory Klyza
and
Stephen C. Trombulak

MIDDLEBURY COLLEGE PRESS
Published by University Press of New England
Hanover and London

We dedicate this book to all of the children of the next seven generations, who will help determine the future of Vermont.

Middlebury College Press

Published by University Press of New England, Hanover, NH 03755

Printed in the United States of America 5 4 3 2 1

CIP data appear at the end of the book

Contents

Illustrations and Tables

Maps

Figures

Tables

Foreword

The Story of Vermont: A Natural and Cultural History inaugurates a new series in Environmental Studies, cosponsored by Middlebury College and the University Press of New England. The books in this series will pursue a bioregional approach to environmental topics. Such an approach emphasizes the continuity between natural history and human history and often seeks to illuminate such connections by focusing closely on particular ecosystems. The inclusiveness found in bioregionalism is one natural outgrowth of the complex environmental history of New England and the Adirondacks. Wilderness and culture have been intricately interwoven in this long-settled region, which has seen a dramatic resurgence of forests and wildlife. The editors of the Middlebury Series believe that a healthy irony can enter into America's environmental discourse through study of this region's turbulent history and surprising present.

The Story of Vermont does not emphasize the state's political boundaries. On the one hand, it strives for the precision of a smaller scale, delineating six main biophysical regions into which Vermont can be divided and within each of them evoking with considerable scientific precision the life cycles and ecology of characteristic organisms. On the other hand, it insists upon larger natural and cultural affinities that link parts of Vermont to the Greater Laurentian Region as a whole, that connect our Champlain Valley with the forests and mountains just across Lake Champlain in New York, and that make the upper Connecticut River Valley of Vermont and New Hampshire a landscape with its own integrity.

In this book, such an insistence upon Vermont's concrete, natural context is reinforced by an emphasis upon the human story as included within and integral to the larger natural history. The narrative of Vermont begins, of course, with geology, and the authors evoke with unusual clarity the tectonic and glacial dynamics that have shaped today's landscape. But as their account arrives at more recent developments—such as the forest history of Vermont since the last glaciation and the related fluctuations of its wildlife—the book also comes to encompass political, agricultural, economic, and technological dimensions of human history. Patterns of settlement and development in Vermont have in many particular ways both reflected and decisively shaped its dynamic landscape.

The dramatic changes that have always characterized Vermont, including the reversals of its recent forest history, have led the authors to conclude by projecting alternate scenarios for the state's future. In the decades immediately ahead, trends in human population and transportation will have vast implications for the nonhuman communities of the state. An important aim of this book, and of the series in environmental studies that it inaugurates, is to foster a more informed sense of the mutual influence and possible balance between people and their landscape. *The Story of Vermont* focuses on just one region of northern New England. In the connections it draws and the questions it raises, however, it suggests a fruitful way of exploring the natural and cultural histories of many other diverse landscapes.

John Elder
Middlebury College

Acknowledgments

This book tells the story of over a billion years of geological, biological, and cultural history. We would never have been able to tell this story without the help of a multitude of people who have generously given their time and expertise to share with us information that they have acquired about the natural and cultural history of Vermont.

The perspective we have taken in telling this story developed over many years and benefited from our interactions with many people, too numerous to list completely. Some of the people who have made the greatest contributions to our educations about the natural and cultural history of Vermont have been Jim Andrews, David Brynn, Diane Burbank, Dave Capen, Charlie Cogbill, Marc DesMeules, John Elder, Brett Engstrom, Steve Faccio, Chris Fichtel, Clay Grove, Charles Johnson, Bill Kilpatrick, Barry and Warren King, Rich Langdon, Marc Lapin, Everett Marshall, Steve Mylon, Steve Parren, Bob Popp, Carl Reidel, Chris Rimmer, and Liz Thompson. All of these people are central to the efforts that are going on throughout the state today to improve general knowledge about Vermont and to create a future that provides for healthy communities, both natural and cultural. They have all given generously of their time to make Vermont a better place to live, and we are extremely grateful for their efforts, even those separate from their contributions to the knowledge base we present in this book.

In addition, a number of people and institutions have been very helpful in finding data and sources. The individuals include Preston Bristow, Vermont Land Trust; Bob DeGeus, Vermont Department of Forests, Parks and Recreation; Nick Forbes, U.S. Army Corps of Engineers; Tom Frieswyk, U.S. Forest Service; David Lamont, Vermont Public Service Department; Ed Leary, Vermont Department of Forests, Parks and Recreation; Jerry McArdle, Vermont Department of Environmental Conservation; Andy Raubvogel, Vermont Agency of Natural Resources; Hans Raum, Middlebury

College library; and Nancy Sherman, Vermont Public Service Department. The institutions include the libraries at Middlebury College and the University of Vermont and the following Vermont agencies and departments: Agency of Commerce and Community Development; Agency of Natural Resources; Department of Agriculture, Food, and Markets; Department of Health; Department of Motor Vehicles; Department of Public Service; and Department of Tourism and Marketing.

A special thank you goes to the people at Northern Cartographic—especially Ed Antczak and Bob Gagliuso. We relied on their experience, expertise, and skill to produce the maps in this book. We are also grateful to Jim Rodda, who prepared figure 1.1, and Dave Guertin, who helped us digitally scan and modify figure 1.2.

We are also grateful to the generations of our students at Middlebury College, who have provided us the opportunity to organize our ideas into a concise story. Good storytelling is good teaching, and good teaching is learned only from the presence of good students. We count ourselves fortunate to have worked with some of the best students on Earth. Their commitment to interdisciplinary thinking, to a curriculum that focuses on the development of a sense of place, and to pushing themselves toward excellence and beyond has been, and continues to be, a source of inspiration for us.

Portions of this book were read and edited by several people, to whom we are forever indebted: Jim Andrews, Ray Coish, Lucy Harding, Charles Johnson, Marc Lapin, Andi Lloyd, Dorothy Mammen, Pat Manley, Sheila McGrory-Klyza, and Liz Thompson. Going even further beyond the call of duty were those who read and commented on the entire book for us: David Brynn, John Elder, Nan Jenks-Jay, Kathy Morse, Carl Reidel, and especially John Davis, who copyedited the entire manuscript. Their suggestions dramatically improved the accuracy and readability of the text. We confess that we did not take all of their suggestions, and any flaws and errors that remain are entirely our responsibility.

We also thank the Trustees of Middlebury College, who helped provide the time and financial support that made the completion of this book possible. Their success at creating an academic environment where both education and scholarship are viewed as integral parts of the liberal-arts tradition has been central to our intellectual development.

Most important, we also thank our families—Sheila McGrory-Klyza, Isabel McGrory-Klyza, Caroline McGrory-Klyza, Dorothy Mammen, Sage Trombulak, and Ian Trombulak—for the moral support and understanding they have given us throughout this entire project. This book is a reflection of but one aspect of our lives; our families are at the very core of everything.

Middlebury, Vermont C.McG.K.
September 1998 S.T.

The Story of Vermont

Introduction

WE BEGIN THIS story from the top of Mount Abraham, the fifth-tallest mountain in Vermont, on a clear June day. At an elevation of just over four thousand feet and with a small alpine meadow at its summit, this vantage point offers us a fine view. We see four different mountain ranges: the Green Mountains extending to the north and south, the Adirondacks to the west, the Taconics to the southwest, and the White Mountains to the northeast. We see Lake Champlain and the Champlain Valley to the west and the Mad River Valley to the east. We also get a clear view of the vegetation patterns of Vermont. To the south, the spine of the Green Mountains appears to be cloaked in green velvet. Much of the Mad River Valley and the ridge to the east are also forested. In the Champlain Valley, though, we see a more significant overlay of human forces at work, with cropland and pasture mixing with forest. Superimposed on these geological and biological layers are the most obvious works of humans; roads, villages, and houses.

A closer look brings human influences into clearer focus. We have arrived at the summit by way of the Long Trail, which follows the spine of the Green Mountains from Quebec to Massachusetts. The forests we see in three of the mountain ranges have been partially protected by conservation: the Green Mountain National Forest in Vermont, the White Mountain National Forest in New Hampshire, and the Adirondack Park in New York. A bit of extra walking to the north would bring us to a chairlift and the Sugarbush ski area. This resort has influenced the landscape of the Mad River Valley, as similar resorts have influenced similar valleys elsewhere in the state. Land trusts have helped to protect farms and forests within our view.

Our goal in this book is to explain what it is we—and you—are seeing from the summit of Mount Abraham or from any other place in Vermont. The landscape we see is the result of three primary forces: geological, biological, and human. The geological forces of mountain building, glaciation, and erosion are responsible for the principal features. Ecological and climatic forces are responsible for the forest types and the myriad plant and animal species that inhabit the landscape. And finally, but by no means least important, human factors have played a significant role in shaping this landscape, especially in the years since large-scale European settlement began in the region. Perhaps most illustrative of these effects is the change in forest cover in Vermont during this period. The percentage of Vermont that is forested went from an estimated 95 percent in 1620, to 25 to 35 percent around 1850 to 1870, to more than 75 percent in the early 1990s.

Our story of the Vermont landscape is based on three major themes. First, landscape history or natural history without humans is incomplete history. Humans have played a major role in shaping the Vermont landscape; indeed, they have been the dominant force over the last 250 years. A key to understanding any landscape is recognizing that natural history and human history are not fundamentally different subjects but, rather, fundamentally related parts of a single, more comprehensive history. Nature and culture have a dynamic relationship in which each continually influences the other. For example, one basic change in human technology—the arrival of the railroad in Vermont—had far-reaching effects on the landscape. The value of land near the tracks and stations increased; hill farms removed from the tracks declined in value and were often abandoned; the number of forest fires increased due to sparks from the trains; there was a tremendous increase in demand for wood for railroad ties and fuel for the trains; plants moved across the landscape via the rail corridors; and dairy farming surged in Vermont because liquid milk could be sold in Boston.

Our second theme is that the landscape histories of particular regions need to be embedded within the context of larger regions. The natural world does not recognize political boundaries. Rather, variation in the distribution of species and ecosystems results from variation in geological, climatic, and biological factors. These variations, mapped on a continental scale, indicate the presence of ecological regions—areas that share common vegetation types, topography, climate, and species compositions. Looking at the natural history of an area in the context of its ecoregion gives us greater insight into how the area came to be the way it is and what ongoing processes continue to shape it. In order to help us achieve this goal, almost all of the maps in this book will cover one of three regions: (1) the Greater Laurentian Region (GLR), which comprises the eastern half of the Laurentian Mixed Forest Province ecological region, including southern Quebec, parts of the maritime provinces, New England, New York, and northern New Jersey and Pennsylvania (this region is more useful for context than New England alone since Vermont borders Quebec and New York); (2) the central portion of the GLR; and (3) Greater Vermont, encompassing those portions of Massachusetts, New Hampshire, New York, and Quebec that adjoin Vermont. Furthermore, these maps were created with a geographic information system (GIS) that allows us to portray data about this landscape with a level of precision never before available in a book of this type. This GIS representation provides a foundation for greater spatial accuracy, which is the basis for a better understanding of the region's natural and cultural history.

Furthermore, political boundaries are dynamic human constructs, and people living within political units are constantly affected by outside forces. Vermont, with its present legal borders, has existed for only slightly over two hundred years, and those boundaries were very much in dispute for many years. Those living in Vermont—and the landscape itself—have been greatly affected by decisions made beyond the state's borders and by forces beyond their control. Over time, such external human forces have had an increasing influence. Farmers of the early eighteenth century—the vast majority of the state's population—were largely self-sufficient and buffered from events beyond Vermont. Today, those living in the state rely on external supplies for almost all of their food and energy and hence have little control over forces related to these necessities. Conversely, however, since Vermont imports most of its food and energy, the state has become significantly reforested.

The third major theme of our story of the Vermont landscape is that the natural world is not described adequately by a simple list of species. Rather, nature is organized into broadly defined ecosystems, each of which can be described as a collection of biotic and abiotic parts and as a set of processes that influence both how the system operates over time and how the parts persist or change. Ignoring the functional aspects of ecosystems can lead to misconceptions that the natural world is static, rather than dynamic, and that the parts of an ecosystem are infinitely removable or interchangeable. These misconceptions have led—and can still lead—to a great deal of ecological destruction.

In order to understand the Vermont landscape, then, we must understand geological, biological, and human forces; we must understand each of these forces in a historical manner; we must understand that each of these forces and the interactions among them are dynamic; and we must understand that each of these forces extends beyond Vermont's borders and that some of each force's greatest effects on Vermont will be due to events and processes that take place beyond the state's borders.

Before moving on to the story of the Vermont landscape, we must present several ground rules and caveats. First, we use current political names to refer to the landscape even before those political names came into being. That is, even though Vermont as we know it has existed for only a little over two hundred years, we use "Vermont" to designate this geographic place when the Abenaki lived there, when it was under a sheet of ice, and when it was part of the creation of the continent more than a billion years ago. Second, we tell the story based on the best available knowledge we have today, knowing full well that the understanding of the past is constantly changing. For instance, anthropologists are continually revising the date for when they believe humans first arrived in North America and the way in which these humans spread across the Americas. Third, we are deliberately painting with a broad brush. We are concerned with overall temporal and spatial patterns and trends through the millions of years of geological history and the thousands of years of postglacial biological and human history of Vermont and over the roughly six million acres that constitute Vermont.

The first six chapters tell the story of Vermont in chronological order. Chapter 1 describes the creation of the landscape that constitutes the Greater Laurentian Region, with special attention to the major mountain-building episodes in the region and the last ice age. Chapter 2 traces the arrival of plants, animals, and, eventually, humans to Vermont following the recession of the last glacier roughly 13,000 years ago. The story of human effects on the Vermont landscape—and how the landscape affected its human occupants—from the arrival of Europeans in the state through the present is the subject of chapters 3 through 6. The next three chapters describe the three major types of ecological communities found in Vermont today—forest, terrestrial open, and aquatic—and how human settlement has affected their function and composition. These chapters are consciously not part of the chronological narrative because we wanted to provide greater detail on these ecological communities and such detail would severely hamper the flow of the story. To put it another way, chapters 3 through 6 present cultural history in a natural context; chapters 7 through 9 present natural history in a cultural context. In our concluding chapter (10), we offer our thoughts on the major patterns and trends that have shaped Vermont's natural history and speculate on the future of the Vermont landscape.

PART I

Setting the Stage for Vermont

1

The Past as Prelude

The Early Evolution of Vermont's Landscape

THE HISTORY OF the place now called Vermont can be traced back thousands of millions of years. The story of how this landscape was created is not simple. It involves the repeated collision of continents, the birth and death of oceans, and the occasional burial of the land under thousands of feet of ice. Although most of the events that shaped the landscape took place in the far distant past, long before humans even evolved, their consequences are felt strongly in the region even today. Vermont's geological history not only defines its physical landscape but delimits its biological and cultural possibilities. Marble quarries in the Taconic Mountains, copper mines in the Connecticut River Valley, granite quarries in the northern Piedmont, ski slopes in the Green Mountains, fertile agricultural land in the Champlain Valley, deep-water fishing for landlocked ocean species in Lake Champlain, a near-complete cover of deciduous hardwood and spruce-fir forests, and scattered sand and gravel deposits that make on-site septic disposal possible: all of these owe their existence to the region's long and complex geological history. This history is essential to understanding the basis for the biological and cultural events that make up much of the story of Vermont.

The Surface of the Earth

Before the story of the early evolution of Vermont's landscape can be told, a few basic concepts about the earth's physical structure must be understood. The entire surface of the earth is dynamic. It is always in motion, being constantly reshaped and rearranged. This motion is a consequence of Earth's internal structure. Its outermost layer, the lithosphere, forms a rigid shell about sixty miles thick over the entire planet (see fig. 1.1). The lithosphere

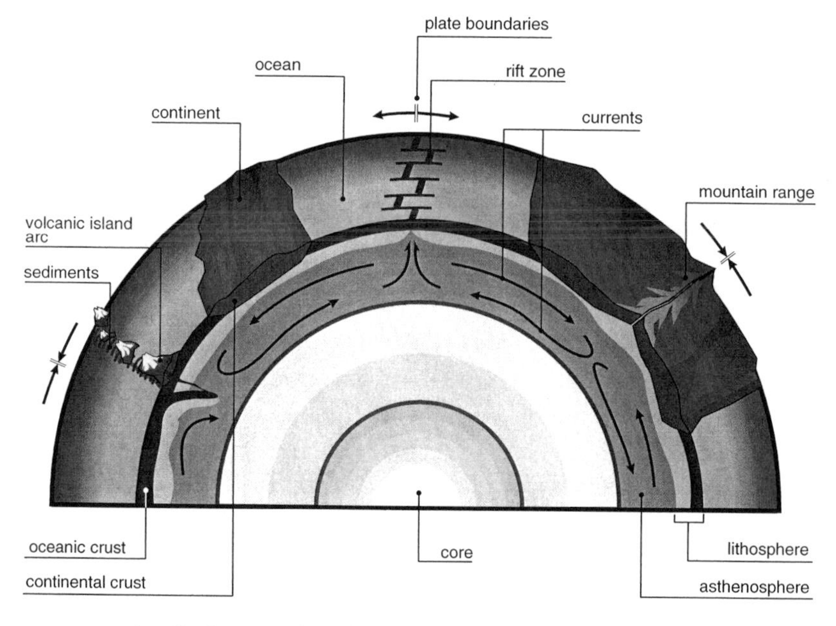

FIGURE 1.1. A stylized cross section of the Earth, showing the major features associated with the movement of plates.

floats on the asthenosphere, a partially melted layer of rock more than 1,700 miles thick. The asthenosphere remains partially molten because of the intense heat in the earth's deep interior. The asthenosphere slowly flows in currents, rising and sinking in many places around the globe. Gravity and the asthenosphere's currents in turn exert forces on the lithosphere above it, breaking it into pieces, or plates, and moving them around.

Each lithospheric plate has an outer crust. In some places, the crust is only about four miles thick and is rich in the heavy elements iron, magnesium, and calcium. In other places, the crust is twenty to forty-five miles thick and has a much smaller percentage of heavy elements. This variation in the thickness and density of the crust results in a planetary surface that is extremely irregular. The thick, low-density crust rises high to form continents, and the thin, high-density crust forms the ocean floors. Currently, the lithosphere is broken into at least thirteen separate plates, six of which include major masses of continental crust. But this has not always been so. The movement of the asthenosphere has caused the plates to break apart and fuse together all throughout Earth's history.

Because the geological forces that shaped the present-day landscape of Vermont operated over a very long period of time, much of the story about how this landscape evolved—a story that can only be pieced together from the rocks and landforms evident today—is not entirely certain. The record of these forces was not preserved everywhere, and the timing of events can only be estimated to within a few tens of millions of years. Although the story that follows is widely accepted as the most likely scenario of geological events in Vermont, the results of future research will almost certainly require changes in many details.

The Earliest Landscape

The stable nucleus of today's North America is a platform of rock that underlies the northern and central portion of the continent. Much of its early history is unknown, but sediments that eroded from ancient mountains on this nucleus contain minerals that are at least 2,700 million years old, almost half as old as the earth itself at 4,600 million years. For clarity, earlier versions of the continent are collectively known as proto–North America. The eastern edge of proto–North America up until about 1,300 million years ago ran roughly from what is now eastern Mexico, across northern Louisiana and Alabama, up through the western Carolinas, northern New Jersey, the Connecticut River Valley, and the Gaspé Peninsula of Quebec. Hence, the region of present-day Vermont was, at that time, a coastal plain.

West of this ancient shore was a broad region known as the Grenville Province, which was formed from the erosion of ancient mountains in the continental interior. These sediments fused to become rock—a process called lithification—and later deformed and recrystalized to form much of the rock underlying eastern North America today. In the southern United States, the Grenville rocks are now almost completely buried under younger rocks and sediment, but they are largely exposed in Labrador, southern Quebec, and the Adirondack Mountains.

Around 1,200 million years ago, proto–North America began to move toward another continent that lay to the east (see fig. 1.2a). As it went, the leading oceanic edge of the adjacent plate subducted under the leading edge of proto–North America. The subducted plate heated up from the increased pressure and partially melted. This magma then rose to the surface through overlying cracks, forming a chain of volcanos along proto–North America's eastern margin. In less than 100 million years, proto–North America collided with the approaching eastern continent, pushing the chain of volcanos, ocean sediments, and surrounding rock upward into a Himalaya-like mountain range of folded, broken, and highly metamorphosed rocks (see fig. 1.2b).

This collision is an example of one process whereby mountains are formed. This process is called orogeny and involves thrusting, folding, and

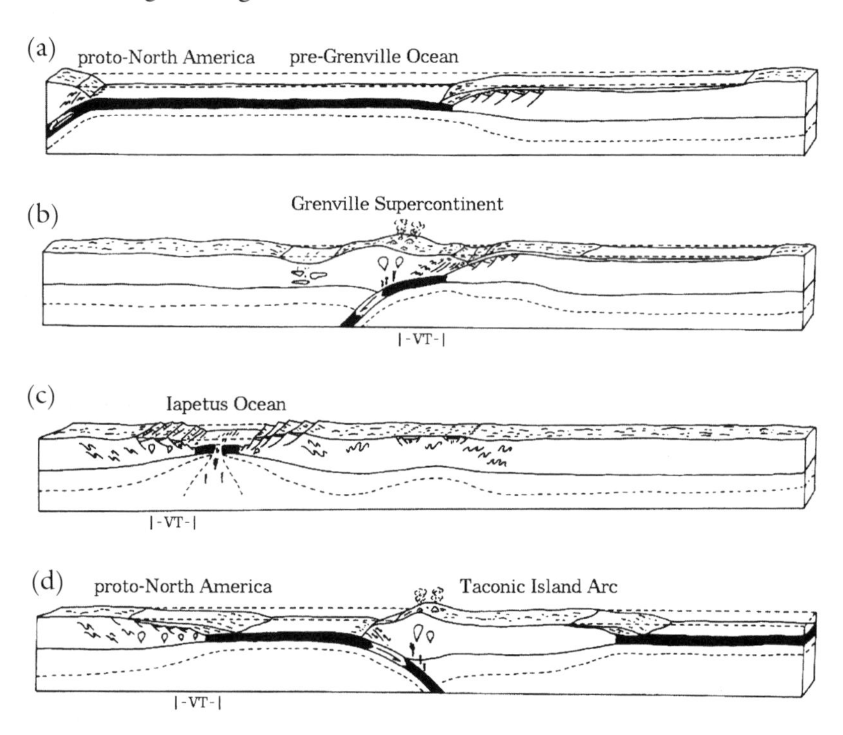

FIGURE 1.2. Evolution of Vermont's landscape. (a) 1,200 million years ago: proto–North America of Grenville Age moves eastward. (b) 1,100 million years ago: Grenville Orogeny and the formation of the Grenville Supercontinent. (c) 575 million years ago: proto–North America moves westward to form the Iapetus Ocean. (d) 500 million years ago: proto–North America moves eastward to create the Taconic Island Arc. (*continued*)

fracturing in the outer and higher layers of the crust. The collision more than 1,100 million years ago, called the Grenville Orogeny, also resulted in the proto–North American plate fusing with the plate to the east to form the Grenville Supercontinent. Indeed, the Grenville Supercontinent was perhaps the fusion of all the continental crust on Earth into a single global continent.

The Grenville Orogeny continued for about 80 million years, but eventually the rate of erosion exceeded the rate of uplift, and the mountain range began to wear away. For 300 million years, until 740 million years ago, the range continued to erode to a flat, sea-level plain. This mountain range itself is now completely gone, but evidence of the collision that formed it is found in the rocks embedded in the cores of the modern Adirondack and Appalachian Mountains, including the Green Mountains, although these ranges themselves formed at later times. Almost all of the surface of the western flank of the Green Mountains from Ripton in central Vermont to

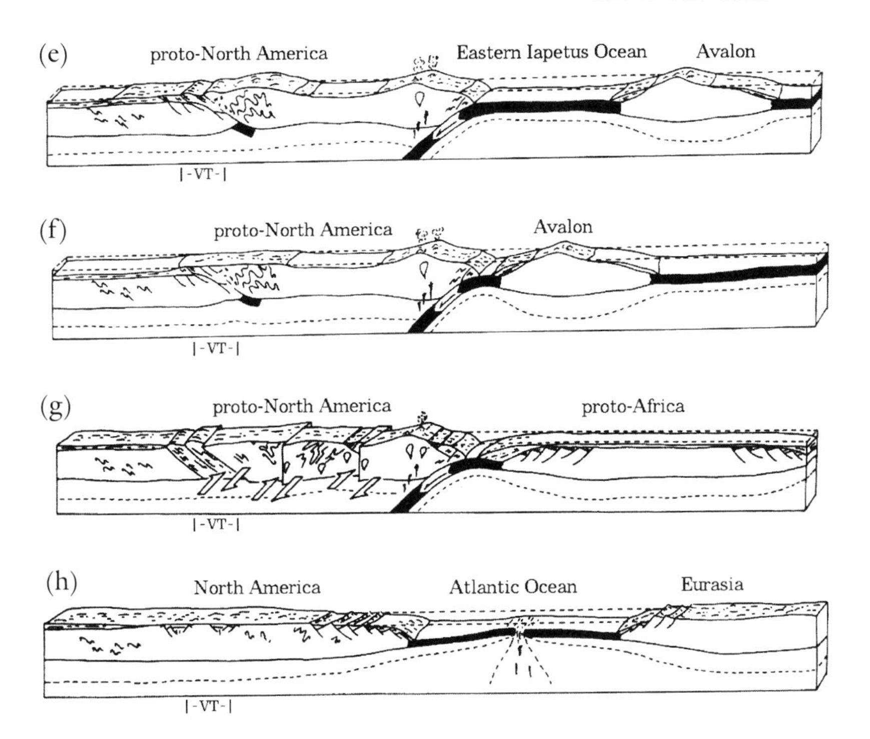

FIGURE 1.2. (*continued*) (e) 450 million years ago: Taconic Orogeny; proto–North America collides with the Taconic Island Arc to form the Taconic and Green Mountains. (f) 400 million years ago: Acadian Orogeny; proto–North America collides with Avalon to add to the Green Mountains and create New England east of the Connecticut River. (g) 320 million years ago: Alleghanian Orogeny; proto–North America collides with proto-Europe and proto-Africa to form the southern Appalachian Mountains and the Pangaea Supercontinent. (h) 180 million years ago: Pangaea breaks up, and the Atlantic Ocean begins to form. Reproduced with permission from the New York Geological Survey/New York State Museum.

just north of Massachusetts is made of rocks that were created during the Grenville Orogeny.

Supercontinents are, by their very nature, unstable. With an uneven distribution of continental and oceanic crust in the lithosphere, forces in the asthenosphere created zones where new crustal material was pushed upward within the supercontinent, and plate separation, or rifting, again occurred. Initial rifting and volcanic activity began, among other places, along the suture where proto–North America had originally fused on its eastern boundary. By 575 million years ago, proto–North America was once again moving westward, leaving a new basin called the Iapetus Ocean opening in its wake (see fig. 1.2c).

During this rifting, much fracturing, or faulting, occurred as large blocks of crust pulled away from each other, allowing magma to fill the cracks and occasionally break the surface. The new eastern edge of proto–North America was once again the site of extensive breaking, volcanic activity, erosion, and sedimentation along the mountain walls and into the ocean. Yet this continental margin was no longer a region of subduction, so volcanic activity eventually stopped, and the east coast of proto–North America became geologically quiet.

The Iapetus Ocean continued to widen for at least 100 million years, reaching a width comparable to the present-day Atlantic Ocean. During this time, sediments washing off the eastern flank of proto–North America accumulated in adjacent coastal waters. These sediments lithified into the diverse sequence of limestones and shales that are seen today in the lowlands and islands of the Champlain Valley and in the Valley of Vermont, the narrow valley that lies between the Green Mountains and Taconic Mountains. Lithified early in the evolutionary history of multicellular life, these rocks reveal a rich diversity of ancient life forms, including algae, sponges, corals, bivalves, nautiluses, and trilobites. Furthermore, at this time proto–North America was situated in tropical water of the Southern Hemisphere, having not yet begun its eventual movement into temperate northern latitudes, so the life forms preserved in these rocks represent very different types of organisms than those present in the region today.

The Taconic Orogeny

Then, about 500 million years ago, the relative direction of the proto–North American plate reversed once again (see fig. 1.2d). As the plate moved eastward, its oceanic part was subducted under its counterpart on the adjacent plate. The resulting melting and volcanic activity created a chain of islands at the leading edge of the overriding plate. This chain, called the Taconic or Bronson Hill Island Arc, probably extended a very long distance through the Iapetus Ocean, from what is today Alabama to Newfoundland, effectively dividing the Iapetus into eastern and western parts.

Subduction and island arc growth continued where proto–North America and the more easterly plate met, until by 450 million years ago the Taconic Island Arc was an immense land mass many miles wide, with high volcanos on its western edge and broad plains of sediment to the east, spreading out into the basin of the eastern Iapetus Ocean. At about this time, the Taconic Island Arc collided and fused with proto–North America (see fig. 1.2e). This collision, the Taconic Orogeny, formed another tall chain of mountains stretching along the entire eastern margin of the continent.

The mountain ranges that formed during the Taconic Orogeny have largely eroded, the sediments mostly carried westward into the shallow sea that at that time filled most of the interior of proto–North America. Remnants of these mountains can still be seen, the two most notable in Vermont being the Taconic Mountains, which run between Vermont, Massachusetts, and Connecticut on the east and New York on the west, and the Green Mountains, which run as a backbone north and south through the center of the state. Material from the arc itself also forms much of the lowlands on the eastern side of the Connecticut River.

The collision of the Taconic Island Arc with proto–North America caused the crust to break into faulted blocks, which overrode each other all along the line of contact. Much of the surrounding rock deformed and recrystallized due to the high temperatures and pressures caused by the collision. This collision is probably the origin of many of the minerals mined in Vermont in historic times, including iron, gypsum, copper, talc, soapstone, and asbestos. The distribution of the precursor minerals in sediments, metamorphic processes, and the infusion of mineral-rich liquids into cracks in the rock led to deposits of particular minerals in locations that today are close to the surface (and are discussed in more detail in chapters 4 and 5). The amount of compression caused by the collision was enormous, and faulting resulted in extensive overlap and jumbling of blocks. Faults that later created the Champlain Valley by allowing large blocks of rock to sink relative to their surroundings, perhaps including the Champlain thrust fault that runs almost the full length of Lake Champlain, formed at this time, as did the fault along which the Saint Lawrence River now flows.

The Acadian Orogeny

The force of the collision between the Taconic Island Arc and proto–North America was so great that the oceanic crust attached on the eastern edge of the Taconic Island Arc, which was now fused with proto–North America, snapped. As proto–North America continued to move eastward, the oceanic crust of the adjacent plate began to subduct under the proto–North American plate. Volcanic activity was thus renewed, but this time it occurred right along the new eastern edge of the enlarged proto–North America. Closure of the Iapetus Ocean continued. Large masses of granite, formed from magma that cooled and lithified underground, rose up, displayed today as granite domes in northeastern Vermont and western New Hampshire.

Sometime during the preceding 200 million years, while the Iapetus Ocean was just forming, a small chunk of crust broke from a continent further to the east, probably proto-Africa, and rafted out into the eastern Iapetus Ocean. After the collision of the Taconic Island Arc with proto–North America and the continued closure of the ocean, this microcontinent, called

Avalon, moved steadily closer to the proto–North American plate. Their collision occurred about 400 million years ago, beginning the Acadian Orogeny (see fig. 1.2f). The effects of this collision were probably less widespread in proto–North America than those of the earlier orogenies because Avalon is thought to have been smaller than other continents or the Taconic Island Arc, but it was sufficient to produce more uplift and folding in the northern portion of the Appalachian Mountains, including the Green Mountains. Avalon fused with proto–North America, adding the land that now makes up eastern New England, including most of the land east of the Connecticut River and the eastern portion of the Maritime Provinces.

The Allegheny Orogeny and the Formation of Pangaea

Following the Acadian Orogeny, subduction continued until about 320 million years ago, when proto–North America collided with proto-Africa (south of what is now Newfoundland) and proto-Europe (along Greenland). The resulting uplift, the Allegheny Orogeny (see fig. 1.2g), had surprisingly little effect in the north but resulted in the formation of the southern Appalachian Mountains, stretching from Alabama to Pennsylvania and New Jersey. Thus, what is today traditionally called the Appalachian Mountains really represents several different mountain chains, formed at different times over a span of at least 130 million years.

The Iapetus Ocean was now, after more than 300 million years, completely closed, and all of the major continents on Earth were once again fused into a single supercontinent, called Pangaea. As happened before to the Grenville Supercontinent, instabilities caused by the unequal distribution of continental crust soon caused rifting to tear the new supercontinent apart. About 180 million years ago, Pangaea broke into northern (Laurasia) and southern (Gondwanaland) portions, roughly along the line of the present-day Caribbean and Mediterranean Seas. Water filled the basin created by the newly formed oceanic crust to form the Tethys Sea.

Soon after, Laurasia and Gondwanaland themselves broke up into smaller continents. North America quickly separated from Eurasia (see fig. 1.2h), and the growing rift zone, the mid-Atlantic Ridge, and the shifting of the continents caused the Tethys Sea to differentiate into the Atlantic Ocean and the Mediterranean Sea.

North America Today

The rift between North America and Europe occurred almost, but not exactly, at the suture that formed between proto–North America, proto-

Africa, and proto-Europe 120 million years earlier. However, a few chunks from both sides of the rift were dragged off as parts of different plates. Norway, Scotland, and northern Ireland, now all parts of the Eurasian plate, were originally part of the North American plate near Greenland. Similarly, what is now northwestern Africa was at one time part of the North American plate near the mid-Atlantic states. Northern Florida, however, is a former piece of Africa wedged in between two chunks of North America.

For the past 180 million years, the Atlantic Ocean has been widening along the mid-Atlantic Ridge. The Appalachian Mountain ranges have steadily eroded, forming broad fans of sediment on both western and eastern flanks, largely burying the surrounding older rocks. Sometime between twenty and forty million years ago, a very small, localized hot spot under the land to the west of Vermont caused the overlying rock to expand and bulge upward, forming the Adirondack Mountains. Much of the overlying rock formed from recent sediments has eroded away, exposing the metamorphosed rock that dates back 1,100 million years to the Grenville Orogeny. Young mountains made from old rock, they are still rising at a rate of between one and two tenths of an inch per year, thirty times faster than they erode away.

After this time, all of the major pieces of land in Vermont and the surrounding region, and their general shapes as seen today, were in place. The formation, transportation, and metamorphism of the rocks that make up Vermont was complete. Mountains had formed, eroded away, and formed again. Yet many of the present-day surface features of the land derive from recent erosional forces, most prominently the presence and passage of a great sheet of ice.

The Recent Ice Age

The geological record indicates that at least four major periods of glacial activity have occurred during Earth's history: 2,300 million years ago, 750 million years ago, 250 million years ago, and throughout recent time, beginning about two million years ago. It is likely that the onset of each glacial age had the same general causes: cooling climates and a massing of continents at high latitudes, which resulted in the steady accumulation of snow on the ground. As the snow piled up year after year, its weight compressed the underlying snow into ice, eventually becoming a thick, heavy mass many thousands of feet thick. Wherever the ice developed on a slope or grew thicker in some areas than in others, its weight was enough to cause it to flow downhill as a glacier.

Erosion and mountain building have all but erased the marks left by the earliest glacial ages, but the present-day surface features in this region

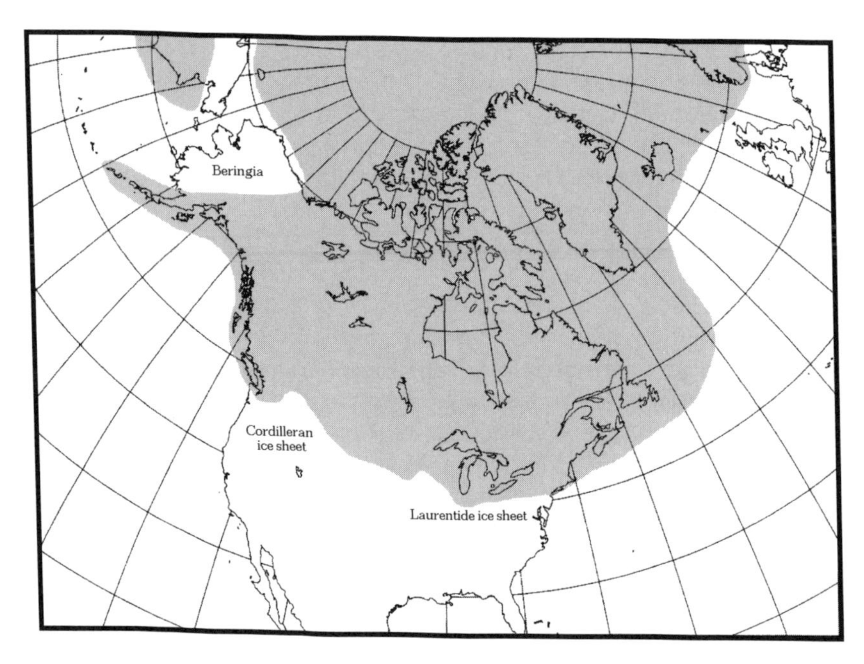

MAP 1.1. The maximum extent of the Laurentide and Cordilleran ice sheets, approximately 20,000 years ago.

strongly reflect the erosion and deposition of the most recent period of glaciation. Around two million years ago, the climate in the Northern Hemisphere cooled, probably due to changes in global ocean and air circulation caused by the creation of Central America and the Himalayan Mountains. These topographic features blocked the spread of warm, tropical water and air into northern latitudes, suddenly making the winters there longer and colder.

Ice sheets—centered over Greenland, central Canada, and the mountains along the Pacific Coast—formed, thickened, and flowed southward. The largest of these was the Laurentide ice sheet, which stretched from the Rocky Mountains in the west to the Atlantic Ocean (see map 1.1). West of the Rocky Mountains lay the Cordilleran ice sheet, stretching to the Pacific Ocean. They flowed southward under the weight of their greatest mass in the north, reaching their furthest extent more than twenty thousand years ago along a line that runs from the present-day Pacific Northwest across the midwestern United States to Long Island. During this time, Vermont was buried under a sheet of ice that in many places was more than one mile thick.

Glacial ice is a powerful force of erosion. As it flows, it scours the land underneath clean of all soil and life down to the bedrock. Where the glaciers

FIGURE 1.3. Camel's Hump in the central Green Mountains. Reproduced with permission from Jonathan Blake.

flow along the sides of mountains, they carve deep U-shaped valleys with relatively steep walls, such as exist today in the Champlain Valley and the Connecticut River Valley.

The flow of ice over mountain peaks scrapes shallow slopes on the up-flow side and steep cliffs on the down-flow side, where entire blocks of rock are plucked off and carried away. This characteristic shape is seen today on Camel's Hump in the central Green Mountains (see fig. 1.3). Boulders are carried along in the moving ice, gouging deep cuts called striations in the underlying bedrock. Striations can still be seen in exposed bedrock throughout the region. The force of the passing glacier is so great that boulders and bedrock alike are eroded into gravel and silt. Much of the rock is ground into a fine powder, called glacial flour, that is eventually carried away in subglacial meltwater and deposited elsewhere.

The most recent glacial age that began two million years ago occurred not as a single advance of a continental glacier but as a series of nineteen or twenty separate glaciations, punctuated by warm interglacial periods. During these warm periods, which were caused by cyclical changes in the seasonal distribution of the sun's energy over the earth, the glacial ice began to melt faster than it flowed forward, and the southern end of the ice sheet shifted northward, giving the impression of a retreat. The most recent glacial advance, the Wisconsin glaciation, lasted for 70,000 years, and was itself characterized by many minor advances and retreats.

About 19,000 years ago, the climate began its most recent warming trend, and the southern edge of the Laurentide ice sheet slowly retreated northward to uncover the northern half of North America. Although the exact timing of the retreat is not certain in all locations, the ice sheet had probably melted off present-day Massachusetts by 15,000 years ago, southern Quebec 12,500 years ago, and northern Quebec 6,000 years ago. A remnant of a continental ice sheet still covers almost all of Greenland and the Arctic islands, so it is likely that the most recent glacial age is not over but merely paused in a temporary interglacial period. One day, as has happened many times before, the cold and glaciers might come again.

The time frame within which geological forces operated to shape Vermont's landscape is immense. The general form of the present-day landscape is a result of many different geological events that took place over a period of time beginning more than a billion years ago and continuing almost up to the present day with the recent ice age. Continental movement and collision created the types of rocks and topography found at any location in Vermont, as well as Vermont's location within North America. Within this broad span of geological time, the landscape supported countless species of plants and animals. Yet for the most part these earliest inhabitants left little record or impact on the present nature of Vermont. The forward passage of the glacier had scoured the earth and pushed all life southward. But with the glacier's retreat, the land was once again exposed, ready for colonization by life and the start of the next chapters in the history of the region, chapters dominated by biological and cultural forces but played out on a landscape shaped by its geological past.

2

The First Colonists

THE RETREAT OF the Laurentide ice sheet uncovered a reshaped landscape. This landscape was once again open to colonization by plants and animals that as a group had been pushed southward 70,000 years before. Many of the species that colonized the region were probably the same ones that had lived there before; 70,000 years is not very long relative to the span of existence for most species. One group of colonists, however, was new to the continent, their ancestors having emigrated to North America from Asia while the Laurentide ice sheet was at its greatest extent. This was the human species, which moved into the region along with numerous other species as the land became free of glacial ice.

Northern New England, with its historic houses and churches, covered bridges, and quaint villages is normally thought of as one of the oldest sections of the country. It was actually among the last places inhabited by humans in the continental United States. Recent findings date the first definite human presence south of the glaciated regions in North America at 12,500 years ago. Most anthropologists think that these humans migrated from ice-free Beringia (the Alaska-Yukon area) through a corridor between the Laurentide and Cordilleran ice sheets, along the eastern front of the Rocky Mountains, or from Beringia along the Pacific coast, moving southward from one ice-free coastal refuge to the next. The descendants of these immigrants eventually made their way east into Vermont.

The Retreat of the Glaciers

As the Laurentide ice sheet retreated and Vermont eventually became ice free about 13,000 years ago, the exposed landscape bore the marks of the recent passage southward of the ice. Yet the landscape was also shaped by at least four forces associated with the retreat of the ice: deposition of glacial debris, formation of glacial dams, changes in sea level, and crustal depression and rebound.

These four forces describe much of the recent geological history of the Vermont region. At its southernmost limit, the Laurentide ice sheet was stationary for many thousands of years. Rocks of all sizes—from silt grains to boulders—that had been carried along by the ice were dumped in long, broad piles called terminal moraines. The Laurentide ice sheet left its largest terminal moraines at the point of its greatest southern extent, forming Long Island, Block Island, Martha's Vineyard, Nantucket Island, and Cape Cod. Morainal debris also accumulated in a number of places in Vermont, especially in low-lying valleys on both the eastern and western flanks of the Green Mountains and in river valleys in the Taconic Mountains and northern Piedmont, as the rate of retreat slowed or even stopped for a period of time.

The ice sheet probably disappeared from the southern border of Vermont by about 15,000 years ago, and completely left Vermont by at least 12,500 years ago. As the glacier retreated, it left behind several prominent geographical features that still influence the region today. As glacier ice melts, the imbedded material is left lying on the ground as unsorted small-sized debris, called glacial till, and individual boulders, called erratics. This blanket of till and erratics covers virtually the entire state except for most of the Champlain Valley, where the later development of lakes buried the debris under lake sediments, and the higher elevations in the Green Mountains, which were less affected by the ice. These boulders and rocks today provide the raw material for the stone fences common throughout Vermont.

Other, more organized depositional features are less common in Vermont than till, but they are found in a few locations in the state. In some places, blocks of ice broke off from the glacier and became buried in outwash. After the blocks melted, a nearly round depression called a kettlehole was formed, later giving rise to some of the bogs found in Vermont. Some debris, particularly smaller-grained sediments, was transported as outwash away from the glacier by meltwater and deposited wherever the force of the stream weakened. Some of these deposits formed broad fans of sediment called kames. Kames, which are distributed widely in Vermont, are found in the Nulhegan Basin in the Northeastern Highlands, the edges of the Valley of Vermont and the Champlain Valley along which the edges of the retreating glacier would have rested, and in high-elevation river valleys throughout the Taconic and southern Green Mountains.

Occasionally, streams that tunneled underneath or through the receding glacier deposited sediments that, after the ice had melted, appeared as rivers of sediment, or eskers. Some of these features are quite large; for example, the Connecticut River Valley esker stretches in sections for twenty-five miles from Windsor, Vermont, to Lyme, New Hampshire. The Passumpsic River Valley esker near Saint Johnsbury runs unbroken for more than twenty miles. Glacial deposits like kames and eskers are the basis for many sand and gravel quarries, as well as sites of groundwater stores for nearby towns.

As the glaciers retreated, they blocked the flow of meltwater to the north like gigantic dams. Two immense lakes built up south of the glacier as it melted off of Vermont and the surrounding region (see map 2.1), along with numerous smaller lakes in high-elevation valleys in the Green Mountains. The first of the large lakes to form was Lake Hitchcock, which built up slowly behind a terminal moraine at Rocky Hill, north of Middletown, Connecticut. As the glacier retreated through Massachusetts, Vermont, and New Hampshire, this lake flooded what is now the Connecticut River Valley.

Although the elevated topography of the surrounding mountains never allowed Lake Hitchcock to be very wide, it grew quite long and sent numerous arms up into tributary valleys. At its maximum extent, it probably stretched as far north as the Canadian border and had an arm up the Second Branch of the White River as far as the Williamstown Gulf. Several surface features in these valleys are thought to have been caused by wave action or sediment deposition at the shore of this lake. Eventually, also about 13,000 years ago, the lake water breached the terminal-moraine dam at Rocky Hill, and Lake Hitchcock drained away to expose the river valleys of today.

As Lake Hitchcock was forming in the Connecticut River Valley, a large lake was also forming from glacial meltwater in the Champlain Valley. The water in this lake flowed northward but was blocked by the ice sheet. Glacial meltwater filled the basin far to the south into the Hudson River Valley, creating Lake Vermont, which at its greatest extent was seven hundred or more feet higher and many times larger than present-day Lake Champlain and reached deep into river valleys in both the Green and Adirondack Mountains. During most of its existence, Lake Vermont drained southward through the Hudson River Valley, which was itself filled in its southern portion by another glacial lake, Lake Albany. Many geological characteristics that were later important to human cultures in the Champlain Valley originated as sediments deposited by surface water flowing out of the Green and Adirondack Mountains into Lake Vermont. These include the valley's excellent agricultural soils and the deposits of kaolin clays, sand, and gravel that have supported mining industries at various times in the past 250 years.

By at least 12,500 years ago, the ice sheet had finally left northern Vermont, and the fresh waters of Lake Vermont began to drain northward into the Saint Lawrence River Valley, eventually dropping the lake to a level that was probably below that of Lake Champlain today. Ten thousand years ear-

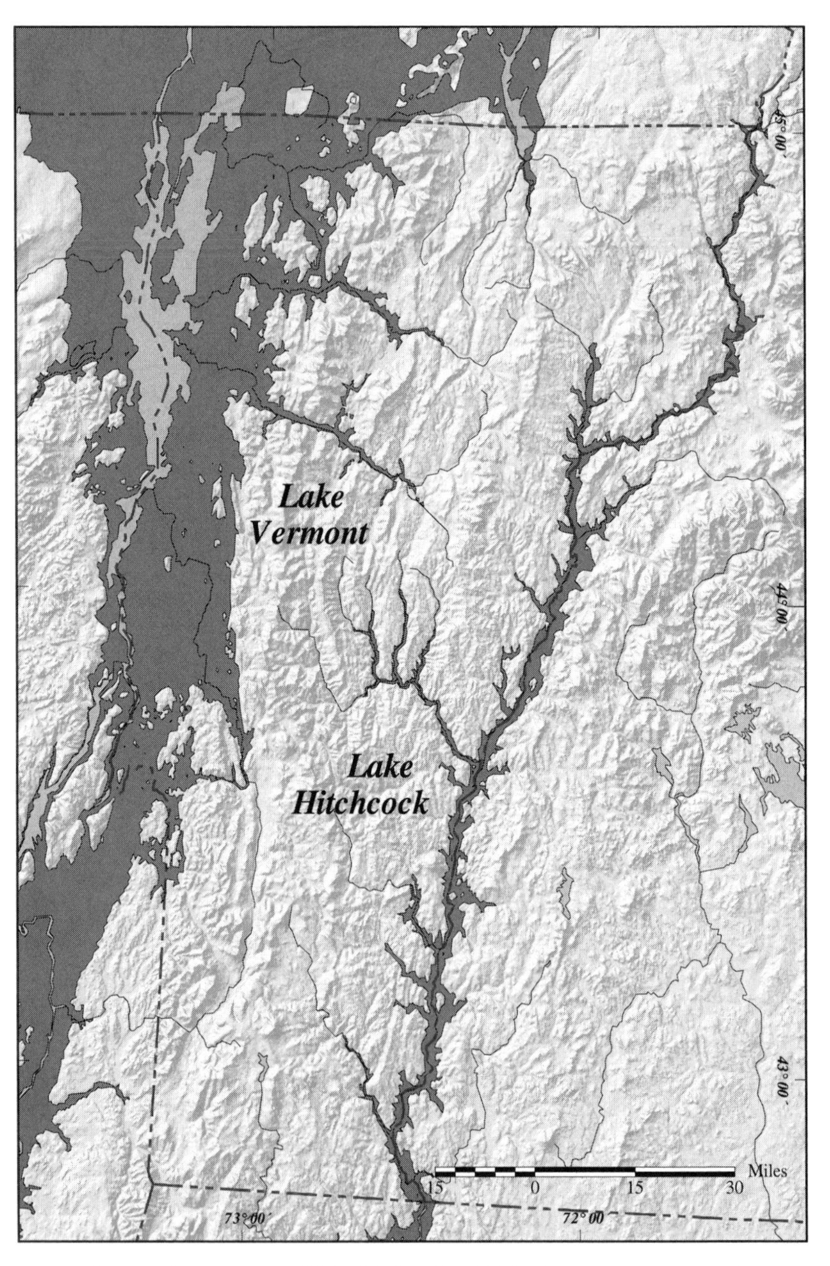

MAP 2.1. The maximum extent of Lake Hitchcock and Lake Vermont, approximately 13,000 years ago. Light area within Lake Vermont indicates the current extent of Lake Champlain.

lier, while glacial ice was widespread, a great deal of the earth's water was locked up in ice, and sea level was at least 250 feet and perhaps 400 feet lower than it is now; the shoreline of eastern North America was as much as one hundred miles to the east of where it is today. But as the ice melted, sea level rose quickly. Once the glacier had melted off of the Saint Lawrence River Valley, seawater flowed up into this valley and into the Ottawa River Valley and the Champlain Valley, creating a new arm of the Atlantic Ocean called the Champlain Sea (see map 2.2). The Champlain Sea was especially extensive because the weight of the recently departed glacier had been enough to depress the continental crust in Vermont an average of five hundred feet, pushing more of the region below sea level.

The Champlain Sea existed for about 2,500 years and, although smaller in area than Lake Vermont, resulted in the deposition of a thick band of marine sediments. Many fossil remains of marine animals, especially clams, dating from this time have been found in the Champlain Valley. The most famous of all these fossils is the nearly complete skeleton of a whale found by workmen building a railroad in Charlotte in 1848.

Eventually, the crust rebounded to higher elevations. The Champlain Valley and the Saint Lawrence River Valley rose above sea level and the saltwater drained from the area. Surface-water runoff filled the basin and created Lake Champlain, today one of the largest lakes in the United States. Lake deposits have covered the marine sediments over most of the Champlain Valley, but the legacy of the Champlain Sea remains: wells bored several hundred feet into the ground in the Champlain Valley can still pump out saltwater. By 10,000 years ago, the landscape of the northern United States and southeastern Canada largely looked as it does today (see map 2.3).

Colonization of the Landscape

While the Laurentide ice sheet was retreating and the geographical features of the region were being reshaped, the biota was changing as well. The plants and animals that had lived south of the glacier's terminus or in coastal ice-free refuges advanced northward. Plants were probably slow to colonize because of the lack of soil; it might have taken many years for the wind to blow enough soil, sand, and glacial flour into the region to support even modest plant growth.

Which species were the first colonists probably varied from place to place, depending on whether the deep soil remained frozen throughout the year and on how wet the ground was. Then, as today, the vegetation of the region existed as a mosaic of community types based on local conditions. In most places, however, the first plants were probably a group of

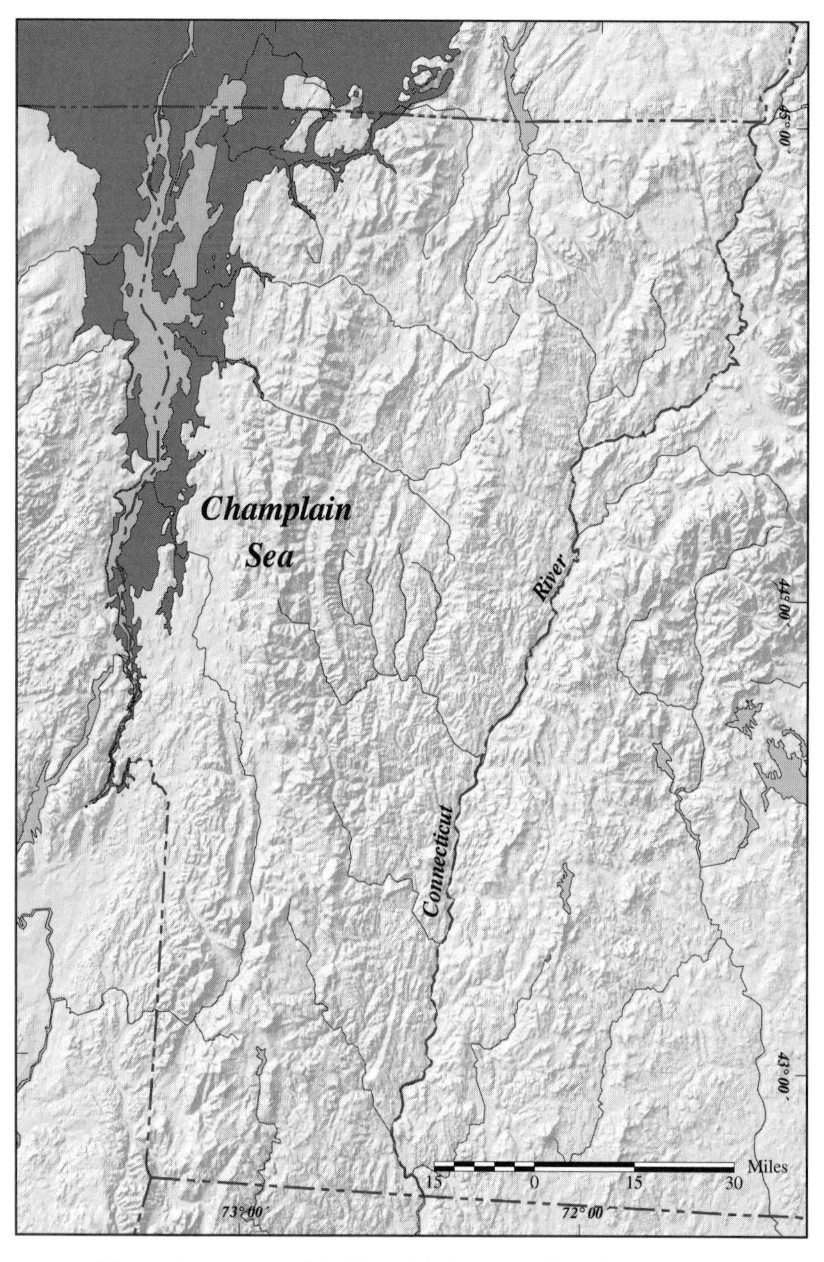

MAP 2.2. The maximum extent of the Champlain Sea, approximately 11,000 years ago. Light area within Champlain Sea indicates the current extent of Lake Champlain.

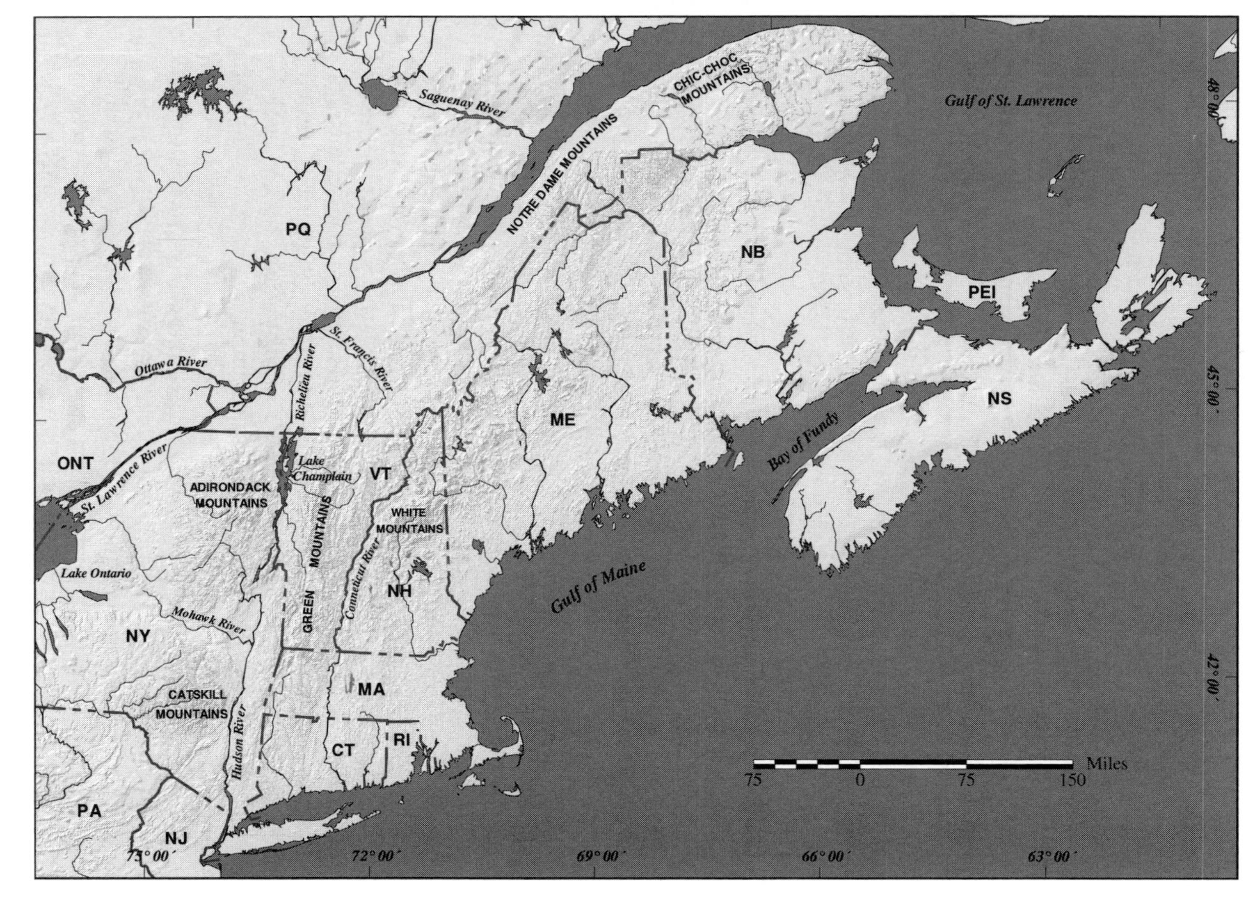

MAP 2.3. The Greater Laurentian Region.

grasses and sedges that today are characteristic of the tundra of northern Canada and Alaska, along with a mix of shrubs such as dwarf willows and alder, as well as ferns, junipers, and club mosses. These low-lying plants provided food and habitat for a host of tundra-dwelling animals including insects, rodents, and large grazing mammals such as woolly mammoths. Teeth of these elephantlike creatures have been found near Mount Holly in the Green Mountains.

Vermont was dominated by a tundra ecosystem for about one thousand years as the climate warmed. The changes in vegetation did not happen at the same time everywhere. Transitions occurred earlier in the south and at low elevations than in the north or at high elevations. In fact, the disappearance of the tundra from Vermont is still not complete: small pockets of tundra vegetation are still present, although precariously, on Mount Mansfield, Camel's Hump, and Mount Abraham, three of the tallest peaks in the Green Mountains.

Beginning more than 11,000 years ago, pollen records indicate that trees began to appear. Black spruce and paper birch grew up in small patches and formed an open woodland that spread throughout the region. Herbs, grasses, and sedges drastically declined as the tree canopy began to close. Numerous woodland species of animals migrated into the area as well. Fossil remains of elk and mastodon have been found in the Champlain Valley, and it seems likely that many other species of large mammal that lived in the eastern part of North America at this time were also present in Vermont. A large number of these spectacular animals suddenly went extinct throughout North America about 10,000 years ago, including the woolly mammoth, mastodon, American lion and cheetah, giant ground sloth, sabertooth, giant short-faced bear, dire wolf, stag moose, and giant beaver (which was the size of a black bear). These extinctions may have been caused by the arrival of humans, discussed below. Many other species that are still part of the region's fauna (or have been extirpated only recently) probably arrived at the same time, including white-tailed deer, black bear, bobcat, timber wolf, mountain lion, wolverine, pine marten, snowshoe hare, and eastern chipmunk.

Eventually, these forests spread and replaced most of the tundra vegetation with a closed forest. In most places, the first closed forest was dominated by red spruce and balsam fir, but open-canopy species such as birch were still able to flourish wherever fire, storms, or insects killed trees to create openings. Spruce and fir are still common throughout much of the higher elevations of Vermont and are the dominant elements of the boreal forests of Canada and Alaska.

By 10,000 years ago, the climate was warm enough to support tree species that during the last great glacial advance had been characteristic of much more southern regions. Eastern white pine, gray and paper birches, and oaks replaced spruce and fir throughout the lower elevations and were the dominant forest type in this region for about three thousand years.

Beginning about eight thousand years ago, several new tree species moved into the region, particularly eastern hemlock, American beech, yellow birch, sugar maple, and American chestnut. Arrival of these species may have coincided with a wetter period, which led to fewer fires and denser forest cover. Roughly five thousand years ago, hemlock suddenly decreased in abundance throughout the Northeast, probably due to a disease, in the same way that American chestnut was struck by a disease and declined earlier this century. Not every hemlock was killed, however, and after a few hundred years they were once again a major part of the low-elevation forests.

The community of plants and animals that existed in this region at any point after the retreat of the ice sheet was not identical to its modern-day counterpart. The tundra of Vermont 10,000 years ago was similar but not identical to the tundra of northern Canada today. The insects that were associated with Vermont's tundra plants 12,000 years ago are today found in the boreal forests, not the tundra, of Canada. Similarly, the boreal forests of Vermont eight thousand years ago were similar but not identical to the boreal forests of central Canada or the spruce-fir forests of high elevations in Vermont today. At that time there was a much greater percentage of deciduous trees such as ash, elm, and oak. What this says is that the composition of natural communities shifts over time in response to local climate and soil conditions, as well as to historical events and variables, such as how long it took a species to migrate northward.

So roughly 4,500 years ago, after the recovery of hemlock, the forests came to resemble fairly closely the forests present in the region at the time of European colonization. A long-term cooling period, which perhaps represents the beginning of the next glaciation, began about 5,500 years ago. It resulted in an expansion of spruce and fir distributions southward and lower down the mountains, but the general distribution of tree species then remains today.

Humans Inhabit Vermont

Four related peoples lived in Vermont prior to the arrival of Europeans: the Paleoindians, the Archaic, the Woodland, and the Abenaki. Each of these different peoples was influenced greatly by the changing landscape. The archeological record suggests that the Abenaki—the people in Vermont when the Europeans arrived—descended from the Archaic people, who in turn perhaps descended from the Paleoindians. The cultural differences

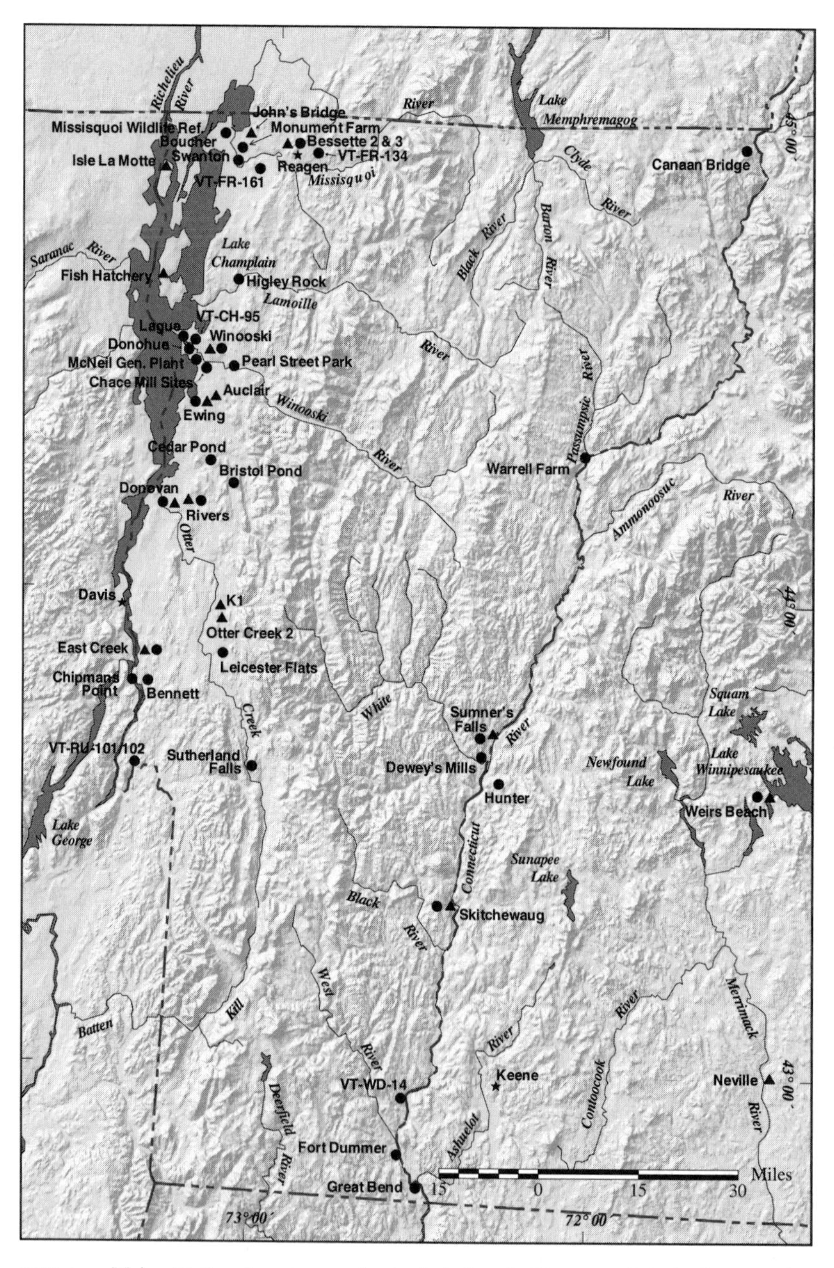

MAP 2.4. Major Native American archeological sites in Vermont: Paleoindians, Archaic, Woodland. *Key:* ★ = Paleoindian; ▲ = Archaic; ● = Woodland.

among these four native groups reflected how each developed and responded to changes in the environment. The Paleoindians inhabited the tundra and open woodland established as the glacial ice sheet receded (see map 2.4). They arrived in Vermont not as a result of a conscious migration but rather through the gradual expansion of their home territory because of changing environmental conditions. Their population was never large, probably less than twenty-five persons per hundred square miles (which would put their Vermont population at less than 2,500). The Paleoindians were a hunting people, preying on the large mammals in the region and the marine creatures in the Champlain Sea, which offered a diverse marine ecosystem rich in mammals, fish, and shellfish. They lived in small groups that moved constantly in search of game.

The Paleoindian way of life in Vermont changed significantly because of the transformation of the marine Champlain Sea to the freshwater Lake Champlain and the change of the terrestrial environment from tundra to forest. Here are two clear examples of the dynamic interaction between humans and their environment. First, the Paleoindians were drawn to a particular landscape and the wildlife it supported following the recession of the glacial ice sheet. Second, when the Champlain Sea and its resources changed, and this tundra shifted to a closed forest, the Paleoindians and their culture changed fundamentally into a new way of life—the Archaic. Furthermore, as mentioned above, between 12,000 and 9,000 years ago, thirty-five to forty large mammals went extinct in North America. Paleontologists are unsure of the cause, but they have developed two main theories. The first, overkill, cites human hunting. The animals had evolved without humans and were unprepared for dealing with these new, spear-throwing hunters, thus making them easy prey. The other main theory is that the extinctions occurred in response to climatic changes at the end of the Wisconsin glaciation. If the extinctions were because of overhunting, this was the first major human effect on the Vermont landscape.

As the landscape changed dramatically, so too did the way of life for the humans living there. The Archaic way of life was based on hunting, fishing, and gathering resources from the new forests that had appeared. Among the resources hunted and gathered were bear, deer, seals, small mammals, turkey and other birds, turtles, fish, berries, nuts, roots, seeds, and other edible plants. The Archaic were a riverine people who probably stayed within specific home watershed territories, smaller than the Paleoindian range, moving on a seasonal basis to take advantage of food resources (see map 2.4). They were, however, influenced by peoples to the south and west regarding the hunting of land animals and by peoples from the east in harvesting lake and river resources. Their population, though unknown, was most likely greater than that of the Paleoindians. This Archaic period came to an end around four thousand years ago, evolving into the Woodland period. This change was due to outside influences and changes in the forest—namely, a decline of hemlocks and rise of hardwood forests. Indeed, by approximately 4,500 years ago, the forests had evolved to the same ecological types that exist in Vermont today. So once again, changes in the landscape led to changes in the culture of those living there.

The Woodland culture developed and reached its peak in the Midwest, beginning 3,500 years ago. With the development of and increasing reliance on native agriculture, this culture was more sedentary. Other major changes during this period included the use of burial ceremonies, pottery, and the bow and arrow, which made for more efficient hunting. These changes gradually made their way to Vermont, with pottery and burial ceremonies arriving approximately 2,500 years ago, followed by the bow and arrow, and eventually agriculture roughly nine hundred years ago. In Vermont, though, these changes were not part of a significantly new native culture, as they were in the Midwest. Rather, they were gradually adopted by and adapted to the existing Archaic culture already in place.

The overall pattern of life continued to center around hunting, fishing, and gathering. Since most important sites from the Woodland period were located along rivers or lakes, fish were clearly a significant food source, and canoes were an important means of travel (see map 2.4). Butternuts were also an important part of Woodland peoples' diets, as they had been for the Archaic peoples. This is an especially interesting detail since the butternut tree in Vermont has recently been hit by a fungus—either an exotic or mutated native one—that could kill off the species in the state. If it does, one of the chief staples of human subsistence for the last four thousand years will disappear from Vermont.

Woodland people tended to live in large villages on rivers, especially near Lake Champlain, with smaller hunting-fishing-gathering camps located along ponds and small streams. The larger villages were occupied seasonally, with the camps used for briefer periods, a pattern of habitation that continued until the arrival of the Europeans in the 1600s. These larger villages and the abundance of middle Woodland sites suggest that the population was now substantially larger than it had been in the late Archaic period.

The earliest evidence of native agriculture in Vermont and for all of northern New England dates to 1100 C.E. at the Skitchewaug site in the Connecticut River Valley. By 1300 C.E., cultivated corn and beans were a major part of the diet of these people and probably contributed to the development of permanent, year-round settlements on higher ground. Such agriculture was not found in the Champlain Valley until after 1400 C.E. Even then, corn, beans, and squash never took on the importance they did in the Connecticut River Valley. There are two major explanations for these differences. The first is based on the different ecological settings of the two areas. The Champlain Valley settlements were farther from the upland food sources, so there was a greater need for mobility than in the Connecticut River Valley settlements. There, a narrower valley and floodplain forced growing populations into earlier and heavier reliance on agriculture. The second explanation suggests that choice was an option in the Champlain Valley. The later adoption of agriculture might indicate that these people were not quickly convinced of agriculture's superiority to hunting, fishing, and gathering. If the resources supplied by their past way of life were in good supply, they had no real need to start farming and to rely heavily on it. When native people adopted agriculture in the Champlain Valley, it fit in well with the existing hunting-fishing-gathering lifestyle; crops were grown on the good soils along rivers and lakes during periods that the people were already living in one place for hunting-fishing-gathering reasons. By the end of the Woodland period, agriculture supplemented hunting-fishing-gathering throughout the region, though in the Champlain Valley it never eclipsed the reliance on wild food. Indeed, by this time, the natives gathered an increased diversity and quantity of wild plant foods.

It should be stressed that from at least as far back as the Woodland period, Lake Champlain was a distinct boundary between native peoples. In the seventeenth century, the lake was a divide between Iroquoian and Abenaki peoples. The Abenaki consisted of the western Abenaki of Vermont and New Hampshire and the eastern Abenaki of Maine (see map 2.5). There were several major bands of Abenaki in Vermont, each associated with major villages on main waterways connected to the Connecticut River or Lake Champlain. Among the chief villages were those at Newbury, Vermont (Kowasek), and Northfield, Massachusetts (Squakheag), on the Connecticut River and at the mouths of Otter Creek, and the Winooski, Lamoille, and Missisquoi Rivers on Lake Champlain. In southwestern Vermont, though, Mahicans rather than Abenaki were the native people present. This region, part of the Hudson River watershed, had been occupied by the Mahicans at least since the middle Woodland period.

The Abenaki way of life, like that of their ancestors, was keenly tuned to the seasons and to place—the keys to survival for any hunter-gatherer-agricultural peoples. Beginning in late April or early May, the Abenaki planted corn, beans, and squash in fertile floodplains near their villages. The Vermont climate, with its late spring and early fall frosts, limited the Abenaki reliance on agriculture. As in the Woodland period, horticulture played an important role but was supplemental to hunting, fishing, and gathering. An additional crop was tobacco, grown primarily for spiritual uses. During the summer, the women generally tended the food crops and gathered berries, fruits, nuts, and other edible vegetation. The nuts and

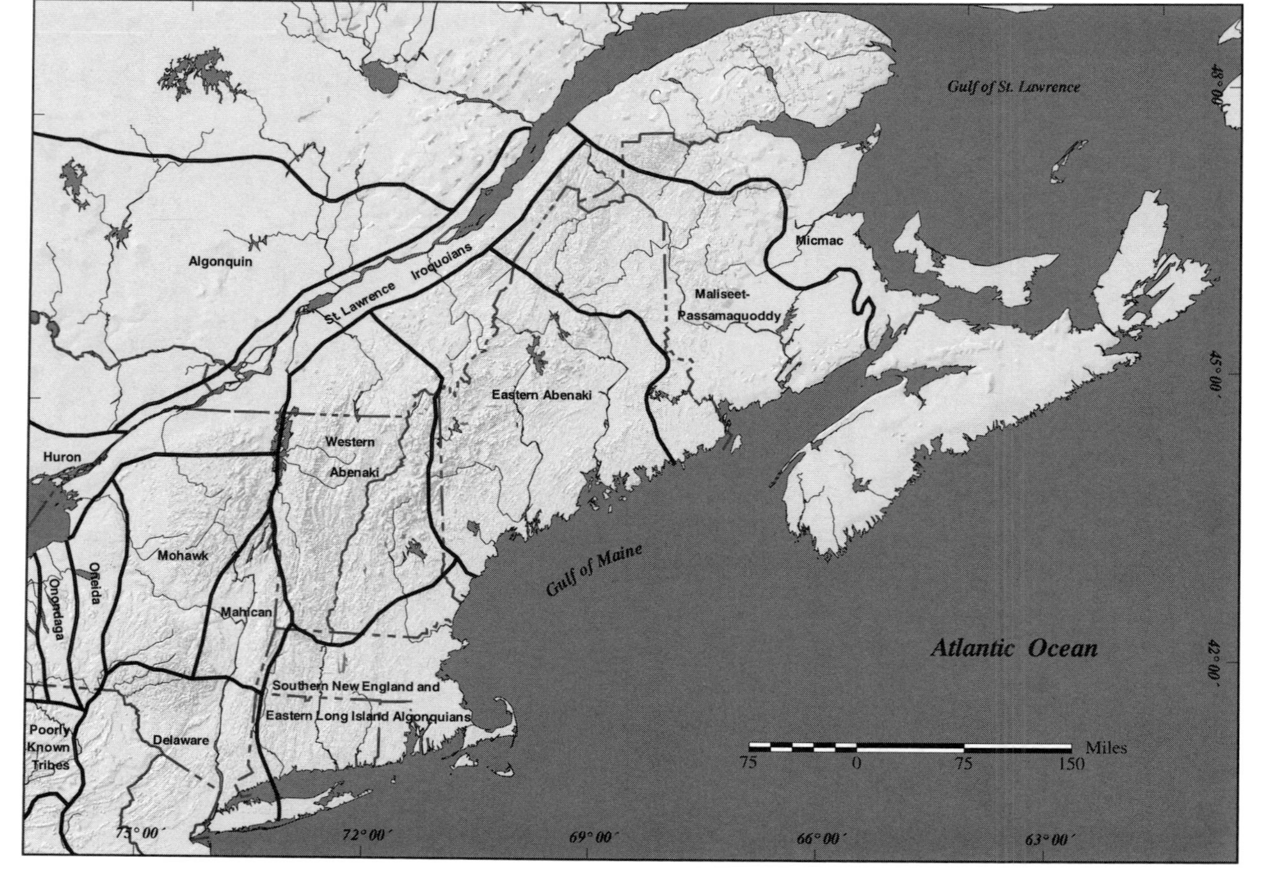

MAP 2.5. Major patterns of Native American occupancy in the Greater Laurentian Region, ca. 1600.

dried berries were stored for winter. The men fished and hunted small mammals, such as muskrats.

In the fall, family bands headed to their hunting territories to hunt for meat and skins, primarily from moose and deer, with some beaver and muskrat as well. Porcupines and spruce grouse were also hunted if food supplies were low. The hunting territories, each of which averaged about twenty square miles, were defined, though they did not have precise boundaries and the Abenaki did not think of property in the same way Europeans did. These hunting territories—the trails and rough boundaries—were watershed based; at their heart were streams connected to larger rivers, with a system of trails feeding to the streams. The territories were divided into quarters by a main trail at a right angle to the stream. Each of these territories was controlled by a family band; others needed permission to hunt or even to travel in this area. There was no buying or selling of land, though. Indeed, the members of the family band had very strong ties to their place. One quarter of the territory was hunted at a time, which reduced the stress on animal populations and the animal wariness that can come from constant hunting. For travel and hunting, lighter birch-bark canoes replaced log dugouts. The birch trees from old-growth forest used to make them were much larger than those found in today's younger forest. In chapter 7, we will describe the characteristics and differences between old-growth and younger forest in greater detail.

The Abenaki spent most of the winter in the larger villages, living off foods prepared earlier, such as dried and smoked meat, corn, and wild plant foods. Midwinter, usually February, was the beginning of the main hunting period, as the stored food supplies began to run low. Hunting—with bow and arrows and spears—was used primarily for moose and, to a lesser extent, deer. Secondary targets, pursued only if encountered on the way, included bear, beaver, mink, muskrat, otter, and porcupine. The Abenaki hunted on snowshoes and sometimes with dogs, giving them advantages over their prey.

As spring came, the Abenaki returned to the larger villages. Maple trees were tapped for sap, and groundnuts and spring greens collected. Bird hunting, including for the now-extinct passenger pigeon, and fishing for shad, salmon, and alewives were very important during this period. The birds and fish constituted nearly all of the Abenaki diet at this time of year, and extra fish were smoked for later use.

The Abenaki population was roughly 25 percent of the level that the land could support, characteristic of hunter-gatherer people throughout the world. This small human population provided a margin of safety against natural fluctuations in food supply. About the time of European arrival (ca. 1600), an estimated 4,000 to 4,200 Abenaki were in the Champlain Valley (a population density of 0.6 people per square mile) and 2,000 to 3,800 in the upper Connecticut River Valley.

The Abenaki of Vermont—indeed, all of the northern New England peoples—differed significantly from the people of southern New England in their relationship with their environment. In the south, farming was dominant, with hunting, fishing, and gathering as the supplemental source of food. This greater reliance on native agriculture also led to greater effects on the landscape in southern New England. More land was cleared for farming, fire was used both to clear land and to keep parts of the forest free from undergrowth to favor certain foods and to make travel easier, and agriculture supported higher population densities. When the Europeans arrived, the native population of New England was estimated at between 70,000 and 100,000, with only 20,000 to 30,000 of those living in northern New England.

Like all humans, the Native Americans who first came to Vermont had an effect on the environment and were in turn affected by it. Prior to the arrival of the Europeans, though, humans' effect on the Vermont landscape was minimal, with the exception of the probable role of the Paleoindians in the extinction of large mammals throughout North America. These Native Americans had low population densities and lived lightly off the land. They hunted at sustainable levels, made minimal use of fire, and did not disturb much land for farming. They followed a cyclical pattern of life tied to seasons and foods and had a very strong relationship to place. In the native worldview, all living things had the ability to act favorably or unfavorably toward humans. This led to a kind of respect for nature, although current society must be careful not to impute a modern environmental consciousness to a people far removed from our time and thinking. In sum, these native peoples had developed a successful way of life in relation to the Vermont environment, one that changed little over four thousand years as the Vermont landscape remained relatively stable. With the European arrival and conquest of the Abenaki (discussed in chapter 3), this way of life was replaced with a completely different mode of interaction with the environment.

The Landscape Today

The geographical landscape of Vermont four hundred years ago, which the Europeans were soon to explore and colonize, was essentially the same as it is today, so it is appropriate at this point to describe the various regions that the Vermont landscape now comprises. There are numerous ways to subdivide a landscape, each based on different combinations of natural landscape characteristics, such as geology, topography, water flow, climate, soil type, and distribution of plants and animals. A landscape can also be either finely or coarsely subdivided, splitting or lumping differences within characteristics to make more or fewer regions.

For the purpose of developing a general appreciation of Vermont's landscape, it is preferable to use a simple classification system that considers broad patterns of geology, topography, climate, and species distributions. This approach divides Vermont into six biophysical regions, but all of them are meaningfully distinct from one another: Green Mountains, Northeastern Highlands, Piedmont, Champlain Valley, Taconic Mountains, and Valley of Vermont. The use of the term "distinct" is potentially misleading because precise borders cannot usually be drawn between one biophysical region and the next. Since such regions are defined by general collections of characteristics, such as "cold climate with a predominance of northern species of trees," the transition from one region to another can take place more or less gradually over a distance of several miles. Despite the fuzziness of the biophysical boundaries, however, each region has a distinct geographical character, and all together they provide a meaningful context for both the natural and cultural history of the Vermont region.

First, running down the midline of the state are the Green Mountains (see map 2.6), a low-relief range that continues as a major feature on the landscape southward into Massachusetts and Connecticut as the Berkshire Hills and northward in sections along the Gaspé Peninsula of Quebec as the Notre Dame Mountains (see map 2.3). The Green Mountain range is on average only about two thousand feet high, but several peaks exceed four thousand feet, namely, Mount Mansfield (4,393 feet), Killington Peak (4,241 feet), Mount Ellen (4,135 feet), Camel's Hump (4,083 feet), Mount Abraham (4,052 feet), and Cutts Peak (4,020 feet). Part of the Appalachian Mountain system, the Green Mountain chain is the dominant landform in the region and divides Vermont into east and west halves, each side with its own distinct character.

The sharp topographic relief of the Green Mountains gives rise to numerous plant-community types (discussed more fully in chapters 7 and 8). Its highest elevations support remnant alpine communities. Below that are spruce-fir forests and, lower still, northern hardwood and hemlock forests. Snowfall can be many feet deep and at high elevations last long into the spring. The Green Mountains themselves vary from south to north, and show a transition from warmer-climate to colder-climate plant communities.

The area east of the Green Mountains, drained largely by the Connecticut River, can be roughly divided into two additional biophysical regions based on general topography, climate, and vegetation. The first of these is defined by the foothills of the White Mountains. Although the bulk of the White Mountains, including its high peaks region, forms the core of neigh-

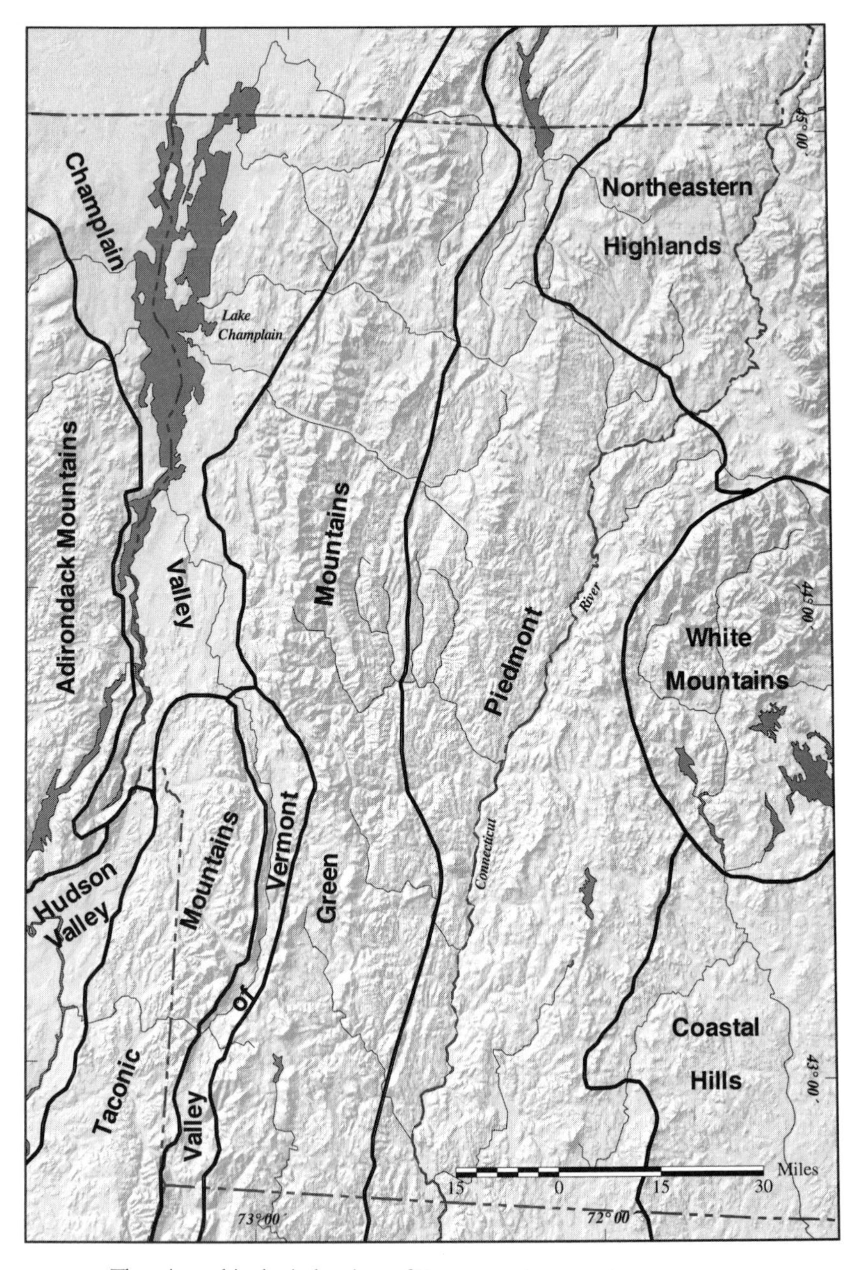

MAP 2.6. The primary biophysical regions of Vermont and surrounding areas.

boring New Hampshire, its northwestern foothills create an upland region in northeastern Vermont. This region is called the Northeastern Highlands or the Upland Plateau (see map 2.6). It is an irregular highland with colder temperatures and greater snowfall than the area further to the south, leading to a greater prevalence of spruce-fir forests and other plants and animals more typical of northern latitudes. The Northeastern Highlands region broadly overlaps with the Northeast Kingdom, a region of Vermont defined by the boundaries of Orleans, Essex, and Caledonia Counties.

However, most of the eastern flank of the Green Mountains slopes gradually down to the Connecticut River to the south of the Northeastern Highlands. These lowland foothills are called the Piedmont (see map 2.6). They are mirrored on the New Hampshire side of the river and characterize the valley all the way into Massachusetts. Here the topography is gentler and lower in elevation than in the Northeastern Highlands; elevations rarely exceed 2,100 feet. These features result in warmer temperatures, less snowfall, and a greater distribution of hardwood-tree species than in the Northeastern Highlands. As with the Green Mountains, the vegetation in the Piedmont varies from south to north.

West of the Green Mountains are Vermont's three other biophysical regions. In the north is the Champlain Valley, a large, broad valley dominated by Lake Champlain, with the greatest surface area of any natural freshwater lake in the United States after the Great Lakes. The Champlain Valley is delineated in the east by the Green Mountains, in the south by the Taconic Mountains, and in the west by the Adirondack Mountains (see map 2.6).

The eastern and western parts of the valley differ from each other. To the west the Adirondack Mountains rise up steeply from Lake Champlain and provide little flat land on the lake's western shore. On the east, however, the land rises only gradually, resulting in a broad low-relief plain with well-developed soils that extends into the foothills of the Green Mountains. The northern end of Lake Champlain extends into Quebec, and the valley opens into a broad plain that stretches all the way to the Saint Lawrence River (see map 2.3).

Lake Champlain covers an area of more than 440 square miles. Its surface is roughly at 100 feet elevation, but the topography of the lake bottom is quite rough. The lake is dotted with numerous islands, some less than an acre large and some that rise as many as 160 feet above the water. The deepest parts of the lake are more than 400 feet deep.

The climate of the Champlain Valley is the mildest anywhere in the state. The Adirondack Mountains to the west block much of the snow that would otherwise move into the region during the winter, and the lake itself moderates the temperatures throughout most of the year. As a result, the valley contains many plant communities that are much more typical of southern

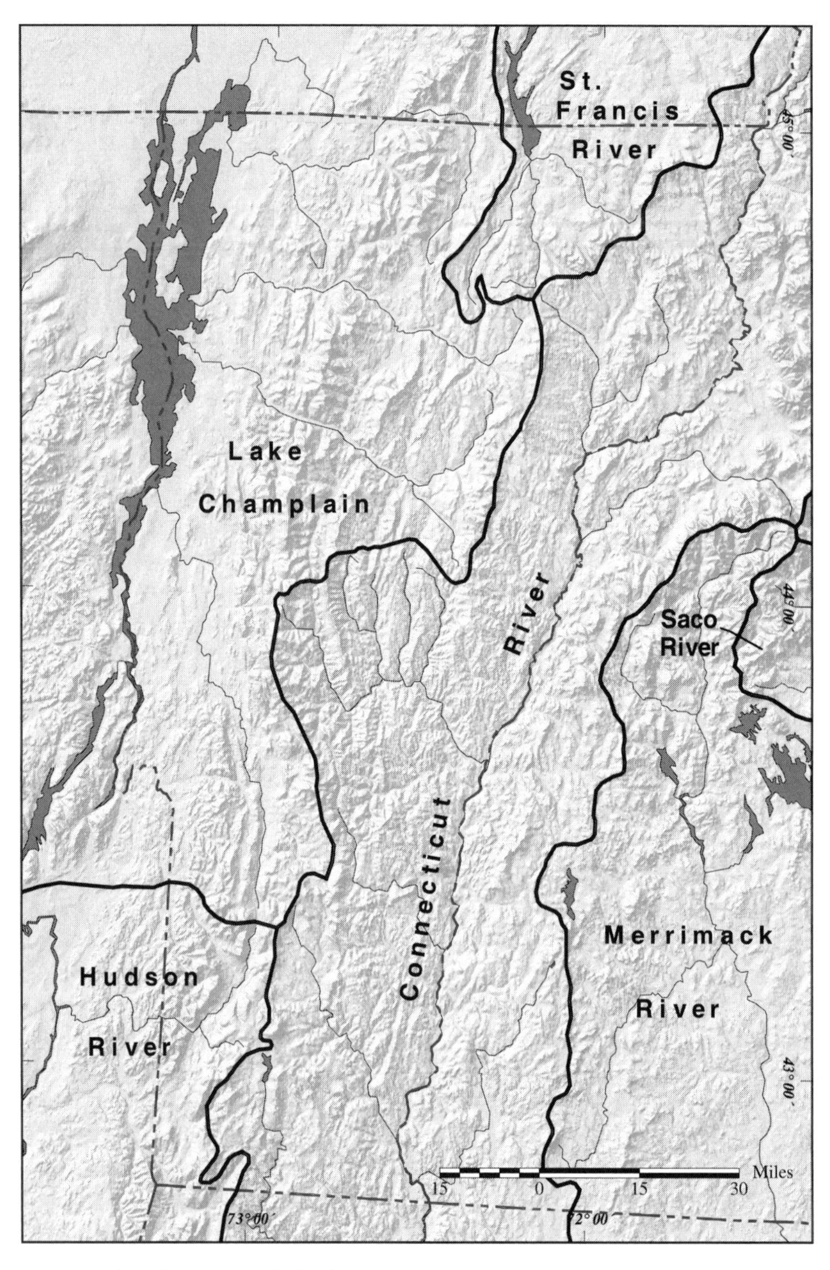

MAP 2.7. The primary watersheds of Vermont and surrounding areas.

New England and maritime regions to the east (the details of which are discussed further in chapter 7).

In southwestern Vermont lie the Taconic Mountains, a small range that runs northward from the Hudson Highlands, along the political border New York shares with Connecticut and Massachusetts, and then up through west-central Vermont as far as Brandon (see map 2.6). The Taconic Mountains lie close to the Green Mountains, in some places separated by no more than half a mile. The Taconics are geologically quite different from the Green Mountains, although they are of similar age, dating back to the Taconic Orogeny 450 million years ago. They are cut more deeply into peaks with sharper ridges and canyons than are the Green Mountains, although their average elevation is lower. The tallest peaks in the Taconic Mountains are Mount Equinox (3,816 feet) and Dorset Peak (3,804 feet), and although there are a few other peaks over three thousand feet tall, most are considerably shorter. The plants and animals here have much in common with those in the Green Mountains, being suited for cold climates and deep snow. The two ranges are so close to one another that many individual animals, particularly bear and deer, travel regularly between the two.

The Green and Taconic Mountain ranges delineate Vermont's sixth geographic region, the narrow Valley of Vermont or Marble Valley (see map 2.6). Topographically, the Valley of Vermont ends just north of the Massachusetts border, but after a small elevation gain it extends geologically southward in the Housatonic River Valley along the west flank of the Berkshire Hills in Massachusetts. The low elevation of the Valley of Vermont generally provides suitable conditions for plants and animals typical of warmer climates. In the north, it is similar to the Champlain Valley, while in the south it is similar to the Hudson River Valley. The bedrock in the Valley of Vermont contains a great deal of limestone and marble, which are rich in calcium, an important plant nutrient. This leads to the presence here of many plant species that predominate where the soil is nutrient rich.

The region's topographic relief creates four great watersheds in Vermont (see map 2.7). The Green Mountains create a major east-west divide down the middle of the state. Two watersheds lie to the east. The Connecticut River watershed, made of the Connecticut River and its many tributaries, particularly the White and Passumpsic Rivers in Vermont, is defined by the Green Mountains on the west and the White Mountains on the north and east. The Connecticut River, which begins in the Connecticut Lakes in northern New Hampshire, flows southward between Vermont and New Hampshire, draining most of eastern Vermont, and passes through western Massachusetts and Connecticut to enter the Atlantic Ocean in Long Island Sound (see map 2.3). The northern part of the Piedmont and the western part of the Northeastern Highlands form a separate smaller watershed, draining northward, mostly through Lake Memphremagog, to the Saint Francis River in Quebec (see map 2.7). This then flows into Lake Saint Pierre and the Saint Lawrence River and then out to the North Atlantic at the Gulf of Saint Lawrence north of the Gaspé Peninsula.

Two other Vermont watersheds lie to the west of the Green Mountains. The largest of these centers on Lake Champlain (see map 2.7). Waters from most of the western Green Mountains, eastern Taconic Mountains, and eastern Adirondack Mountains flow via numerous rivers and streams into Lake Champlain. One of them, the Otter Creek, originates in the Green Mountain above Manchester in the Valley of Vermont and flows northward between the Taconic and Green Mountains for almost seventy-five miles before entering Lake Champlain near Ferrisburg, making it one of the longest north-flowing rivers in the United States. The Winooski, Lamoille, and Missisquoi Rivers also form immense subsidiary watersheds with extensive networks of tributaries that cross the northern portion of the Green Mountains from east to west. Lake Champlain then drains northward through the Richelieu River into the Saint Lawrence River just east of Montreal in Quebec and from there out to the North Atlantic. The fourth watershed is formed by a slight rise in the Valley of Vermont near Manchester (see map 2.7). North of this rise, all water flows into Lake Champlain via the Otter Creek. To the south, water drains through valleys in the Taconic Mountains into the Hudson River, primarily via the Batten Kill and Wallomsac River and out to the Atlantic Ocean south of Long Island.

At no place do the political borders of Vermont completely encompass any unique geographical region. The political entity called Vermont lies at the intersection of many geographical features whose influence stretches far beyond the state's borders. The Taconic and Green Mountains, Valley of Vermont, and Piedmont all link Vermont with southern New England and New York, providing natural corridors for movement of plants, animals, and humans. The Champlain Valley is defined equally by mountain ranges in New York and Vermont, and the flow of its water extends its influence northward into Quebec. The high elevations of the Green Mountains and Northeastern Highlands provide habitat for many species whose ranges are primarily in central and northern Canada. The Northeastern Highlands, Piedmont, and Connecticut River link Vermont eastward with New Hampshire. The political borders of Vermont today, therefore, have little ecological or geological relevance.

The period from approximately fifteen thousand years ago to four hundred years ago was a time of major biotic colonization in Vermont. As the ice sheet retreated and soil developed, plants and animals moved into the region. These living things interacted with each other and the geologic, hydrologic, and atmospheric conditions to create a set of dynamic ecosystems. One of these species—humans—was to come to play a major role in affecting this landscape. Although over the next four hundred years humans would not affect the Vermont landscape to the same degree as the ice sheet, they would have fast, far-reaching, and profound effects.

PART II

The Recent Landscape History of Vermont

3

European Settlement and the Founding of Vermont

THE FIRST EUROPEAN known to have visited Vermont was French explorer Samuel de Champlain, who spent a few weeks in June 1609 on the lake that now bears his name. The first European contact with the Abenaki in Vermont was probably in 1615, when a French missionary visited four communities in the Champlain Valley, but this contact was not renewed until 1642. A few points about European actions and settlement in Vermont should be made at the outset. First, Vermont came under French influence from the north and British influence from the south. Second, as will be described, European traders, the military, and settlers were slow to come to Vermont. Not until 1763 did large numbers of Europeans begin to move into the area.

Sporadic forays into Vermont by the French were designed to keep influence over the area in order to control the fur trade with the Abenaki. They had no desire to colonize Vermont, or any of North America, in the beginning. They established Fort Saint Anne on Isle La Motte in 1666, but it was not maintained for long. A French mission on the east shore of Lake Champlain followed in 1682, but it too was not long occupied. French activity did not pick up again until the 1730s, when the French tried to ward off the British, who were coming from the south, and maintain their claims to the Champlain Valley. They built Fort Saint Frederic at Crown Point (in New York), founded settlements at Alburg and Chimney Point, made a few land grants in the area, and in 1744 established a mission at Missisquoi. In the end, none of these actions mattered after French Canada fell to the British in 1760.

The British route into Vermont was up the Connecticut River Valley from Massachusetts. Like the French, the British were at first interested in the fur trade with the Abenaki. But soon, British settlers wanted the fertile land of the Connecticut River Valley. Fort Dummer was built in 1724, near Brattleboro, to protect new settlers in southeastern Vermont. This fort became the first permanent European settlement in Vermont. Standing in the way of further European settlement were the people who already lived there: the Abenaki.

Native-European Interaction

The arrival of the colonists in New England led to what environmental historian Carolyn Merchant has called an ecological revolution. This colonial ecological revolution resulted in the replacement of the native peoples with Europeans as the dominant humans in the region, as well as a fundamental shift in the ecosystems of the region and in how humans interacted with these natural systems. The Europeans greatly altered the existing landscape, brought new animals and plants, and fully replaced the human way of life that had developed in Vermont over the previous 10,000 years.

The arrival of Europeans in Vermont had disastrous consequences on the Abenaki. They were affected in three main ways: by disease, by the fur trade, and by warfare. In 1616 and 1617, a devastating epidemic of a European disease previously unknown to the natives of New England killed them off by the thousands. This plague affected the peoples of southern New England and the eastern Abenaki, but it is unknown, though likely, whether the western Abenaki were also affected. Native resistance to arriving Europeans lessened because the native population was greatly reduced and native peoples subsequently needed less land and were not as strong in fighting the newcomers. Since the Vermont Abenaki had late contact with the Europeans, the diseases probably made their way to Vermont via other natives. A smallpox outbreak in 1633 in the upper northeast began in the upper Connecticut River Valley and had very significant effects on the Abenaki of that region, killing thousands of them. Further death from diseases, though not epidemics, continued through the mid-1700s, when the number of casualties started to decline due to the drastically reduced size of the native populations. By 1763, when the European influx into Vermont began in earnest, there were not many natives left. Indeed, experts have argued that within one hundred years of first contact with Europeans, native populations had declined by 90 percent. In Vermont, however, the population was boosted somewhat by native refugees from other parts of New England who joined the Abenaki in order to escape the British.

Those natives who survived the new European diseases did not escape other fundamental changes in their way of life. European demand for furs was a key aspect of early native-European interaction. The French established trading posts in the Saint Lawrence–Great Lakes region. The Dutch controlled the trade in the Hudson and Connecticut River Valleys before the English took over in 1664. Throughout the fur-trade period, the Abenaki traded with both the French to the north and the British to the south; the first record of Vermont Abenaki involvement dated from 1648 in southeastern Vermont. Beaver was far and away the most commercially important species taken, though fisher, fox, lynx, marten, mink, moose, muskrat, river otter, and raccoon were also taken for European markets. Bear, deer, and wolf were killed for the local colonial market. The peak of the fur trade in the Connecticut River Valley was in the mid-1650s. Overtrapping, especially of beaver, led to the decline that followed. Since they did not have enough pelts to trade the Abenaki started to trade land to the British for goods. By 1675, with few pelts and no land along the Connecticut River in southern Vermont, the Abenaki were of no further use to the colonists. By 1800, bear, deer, elk, and lynx had disappeared in Vermont south of Middlebury.

Natives traded furs primarily for manufactured cloth and clothing, which replaced the animal hides they had used, and wampum, strings of shell beads from coastal natives that were used as a kind of currency. The wampum was used for the payment of tribute to other tribes, which became more important as the fur trade expanded. Other items that the natives traded furs for were brass, glass bottles, iron goods, kaolin pipes, mirrors, and, from the French, muskets, lead shot, and powder. The fur trade drastically changed natives' incentives to kill these animals, shifting native economies from subsistence ones to commercial ones. As furs took on more value, the boundaries of hunting territories and the rights of those to hunt in them became more defined. This trade with another culture also led to the loss of native technology and worldview. It became harder for the Abenaki to provide their own basic material and spiritual needs. What had begun as the absorption and integration of European goods into Abenaki culture in the mid-1600s had evolved into a dependency on those goods and the culture from which they came.

A dramatic change in the nature and amount of warfare that arrived with the Europeans was the final major force affecting the Abenaki. The Abenaki were no strangers to war prior to the Europeans, having fought neighboring Mohawks since 1570. The major causes for the increased warfare, though, were the fur trade and British-French power struggles. In these latter battles, the Abenaki allied themselves with the French. The warfare was most intense in periods between 1675 and 1760, as the Abenaki fought the British for survival, fearing the fate of southern New England Indians, who had been all but killed off by the British by 1675. These wars had the effect of virtually stopping European settlement in Vermont. The chief battleground was in the Connecticut River Valley, at first near Northfield, Massachusetts. This settlement was attacked repeatedly by the Abenaki and then abandoned by the British. The British always returned, though, with settlers moving farther up the valley. In 1724, the British built Fort Dummer, near Brattleboro, to offer increased protection for settlers heading north; and in 1760, French Quebec fell to the British, leaving the Abenaki without their French allies. It should be stressed that these wars did not have the same devastating effects on the landscape as have modern wars, such as World War II or the Vietnam War. Despite the nearly one hundred years of fighting, the natural communities of Vermont were largely unaffected.

With the Treaty of Paris in 1763 settling this round of British-French wars, British settlers began to surge into Vermont. Lands south of the Missisquoi and the upper Connecticut River Valley were devoid of natives, so there was no human conflict. For the Abenaki in northern Vermont, however, the British ruled that all land south of the forty-fifth parallel (the current Canada-Vermont border) belonged to the king, thus the remaining natives had no claim to it. Nevertheless, the Abenaki were still on their lands when the American Revolution began. Following the Revolution, the Allen family (about which more below) sought to gain control of these remaining lands and have the Abenaki removed. The Abenaki remained the majority into the 1790s at Missisquoi, but by 1800 most had left the village for Canada, and some for poorer lands nearby. Except for this small population, the Abenaki had been replaced by the Europeans as the inhabitants of Vermont.

As the colonists replaced the native peoples throughout New England and Vermont, significant "changes in the land" took place, to use the title of environmental historian William Cronon's fascinating examination of the ecological effects caused by this transition. Upon arriving from Europe, the colonists were especially struck by the incredible abundance of plant and animal life. This was in stark contrast to Europe, and it contributed to the colonists seeing this life as commodities, as natural resources to be exploited. One of the biggest conceptual clashes between the natives and colonists concerned how one owned or managed land. The Abenaki relied on mobility to take advantage of wild resources, and they had hunting territories throughout their lands. The British, however, viewed one's relationship to the land as fixed—not mobile—and brought with them the idea of private land ownership. The British claim to the lands of New England was based on John Cabot's discovery of them, the failure of the natives to subdue and improve the land (that is, clear it and engage in agriculture), and because it was the first Christian monarchy to establish colonies there. For colonists to acquire these lands, they needed a grant from the Crown, or else they needed to buy the lands from someone who had such a grant. The lack of native agriculture—especially in Vermont—meant to the British that these lands were virtually free for the taking. This European conception of property led to fundamental changes in the landscape.

The forests, which had already lost much of their large mammal population because of the fur trade and hunting for meat, changed further at the hands of the Europeans, who cleared them for agriculture. The forests were clear-cut or the trees girdled and burned; there was usually no market for the wood and it had to be disposed of. If there was local demand, charcoal, lumber, and potash were sold. The forests were also cut for firewood and lumber. Certain species were especially valuable—white pines for ship masts (reserved throughout New England for the British Navy since 1711 under the Broad Arrow policy), northern white cedar and white oak for construction, and hickory for firewood—which led to changes in forest species composition. The removal of the forests had further effects. Microclimates changed without the forest cover, leading to hotter summers and colder winters. Flooding and erosion increased as well. This relationship of forest cover to flooding was crucial in the establishment of the Green Mountain National Forest in the 1920s (discussed in chapter 5). In retrospect, even if the beaver, deer, and bear had not been extirpated by hunting, these species might well have been eliminated anyway due to the loss of habitat in the wake of European settlement. We discuss the details of forest community and structure more fully in chapter 7.

Agriculture made the greatest change to existing ecosystems. It led to what Cronon calls "a world of fields and fences." The imported European agriculture had a number of characteristics that distinguished it from the native agriculture of New England. The Europeans brought domesticated animals: cows, oxen, and pigs. These grazing animals required pasture, which led to more forest clearing. They also required fencing, either to protect crops or to keep the animals contained. The oxen allowed the use of plows to cultivate more land. In order to protect these domestic animals from predators, especially the wolf, the colonists used bounties, employed special hunters, and destroyed wolf habitats. The colonists were largely successful, as the wolf was gradually exterminated from New England, though it managed to hang on in Vermont until 1900. The colonists brought their favored European crops to New England, such as wheat, clover, and grass. Although these plants were generally confined to cultivated fields, the colonists also brought a host of plants that did not stay put: weeds. As we will discuss in chapter 8, these exotic plants—and later animals as well—have played a major role in altering Vermont ecosystems. Furthermore, a system of roads was developed to connect these farms, replacing the native system of canoes and trails.

Three final characteristics of European agriculture further fueled the changes in the New England landscape. The colonial farmers were not engaged in a purely subsistence way of life. Many were interested in producing surplus goods to trade for profit or other items. Yet even those who were less interested in marketing their goods needed to produce beyond the subsistence level in order to generate funds to pay taxes. The colonists' reliance on monoculture and livestock grazing led to erosion and soil exhaustion. This meant that after a number of years, colonial farmers abandoned their farms and moved on to fresh, fertile land, necessitating the clearing of yet more forestland. And finally, the sheer scope of the colonial agricultural enterprise meant that the changes it spawned would dwarf any changes caused by native agriculture, especially in Vermont, where such native undertakings were of minimal size.

Overall, when the British replaced the Abenaki as the primary inhabitants, an ecological revolution took place in Vermont that had a number of significant effects on the landscape. The minimal effect on the land of perhaps eight thousand Abenaki farming, fishing, gathering, hunting, and using fire modestly was overshadowed by the activities of the colonists, who would number over 150,000 by 1800. Growing European populations, combined with introduced disease organisms, domesticated animals, grain production, and vast forest clearing, changed the Vermont environment in ways beyond the means and vision of the Abenaki. But there was more. The Abenaki economy had been almost exclusively a local subsistence one, with some trade among neighboring tribes. With the arrival and dominance of the British, the economy in Vermont and in the rest of New England became part of an emerging capitalist system centered on the North Atlantic: Beaver populations were now subject to the whims of fashion in Europe; an expanding British Navy meant more tall, straight white pines cut for ship masts. The Vermont landscape became a set of natural resources for a population much larger than those 150,000 people in the state. In the words of Cronon, "Economic and ecological imperialisms reinforced each other." Forevermore, the nature of human-environment interaction in Vermont was to be influenced profoundly by what happened beyond its borders.

Colonial Vermont

Following the defeat of the French and Abenaki in 1760, the path was clear for British colonists to move into Vermont. Furthermore, the completion of the Crown Point Military Road from Charlestown, New Hampshire, across Vermont to Fort Ticonderoga and Crown Point, New York, around the same time made access to the region easier. This region was a frontier at this time, like Kentucky and Tennessee to the southwest. Furthermore, it was not known as Vermont. Indeed, until 1791 the major issue confronting this land was who had authority over it: New York, New Hampshire, or the new governmental entity of Vermont, formed in 1777. The problem in establishing Vermont's borders—in a sense, of defining what Vermont was and is—arose from confusion over royal grants establishing Massachusetts, New Hampshire, and New York, and later over the boundary between these colonies and British Canada (see map 3.1). So, if history had taken another turn, the natural history of Vermont would simply be subsumed under the natural history of New York or New Hampshire or Massachusetts or Canada. The decisions made in this period determined the borders for our study and defined the political extent of modern Vermont.

Although the main conflict over who controlled Vermont was waged between New York and New Hampshire beginning in the 1750s, there was an earlier dispute over the southern border with Massachusetts in the 1730s. Massachusetts claimed the southern part of Vermont, up to approximately White River Junction. New Hampshire disagreed, arguing for a border further to the south, at a latitude three miles north of where the Merrimack River entered the ocean (roughly at the level of Brattleboro). This disagreement came to a head in 1735 when Massachusetts laid out and granted the current towns of Rockingham and Westminster in Vermont. New Hampshire appealed to the king to settle this boundary dispute. By 1740, the current boundary between Massachusetts and New Hampshire and Vermont was established, a few miles further south than New Hampshire had requested. Massachusetts was forced to surrender its land claims. It should be recalled that settlement in the central Connecticut River Valley was sporadic at this time due to the continuing wars pitting the Abenaki and French against the British.

What the King in Council did not settle in the Massachusetts–New Hampshire dispute, however, was whether the land claimed by Massachusetts in Vermont was actually under the authority of New York, the eastern boundary of which had been established as the Connecticut River in the 1664 royal order establishing the colony. This boundary continued to be ignored by New Hampshire, which claimed a border twenty miles east of the Hudson River, a continuation of New York's border with Connecticut and Massachusetts. New Hampshire's governor, Benning Wentworth, started a series of events that laid the foundation for an independent Vermont when he began to issue charters and make grants of towns in this territory, the first to Bennington in 1749. Prior to the beginning of the French and Indian War in 1754, Wentworth granted sixteen new towns, mostly in southern Vermont. Much of Wentworth's action was driven by land speculation: He received parcels of land in each of these new towns, and if New Hampshire could gain authority over the lands, he could sell them for a healthy profit.

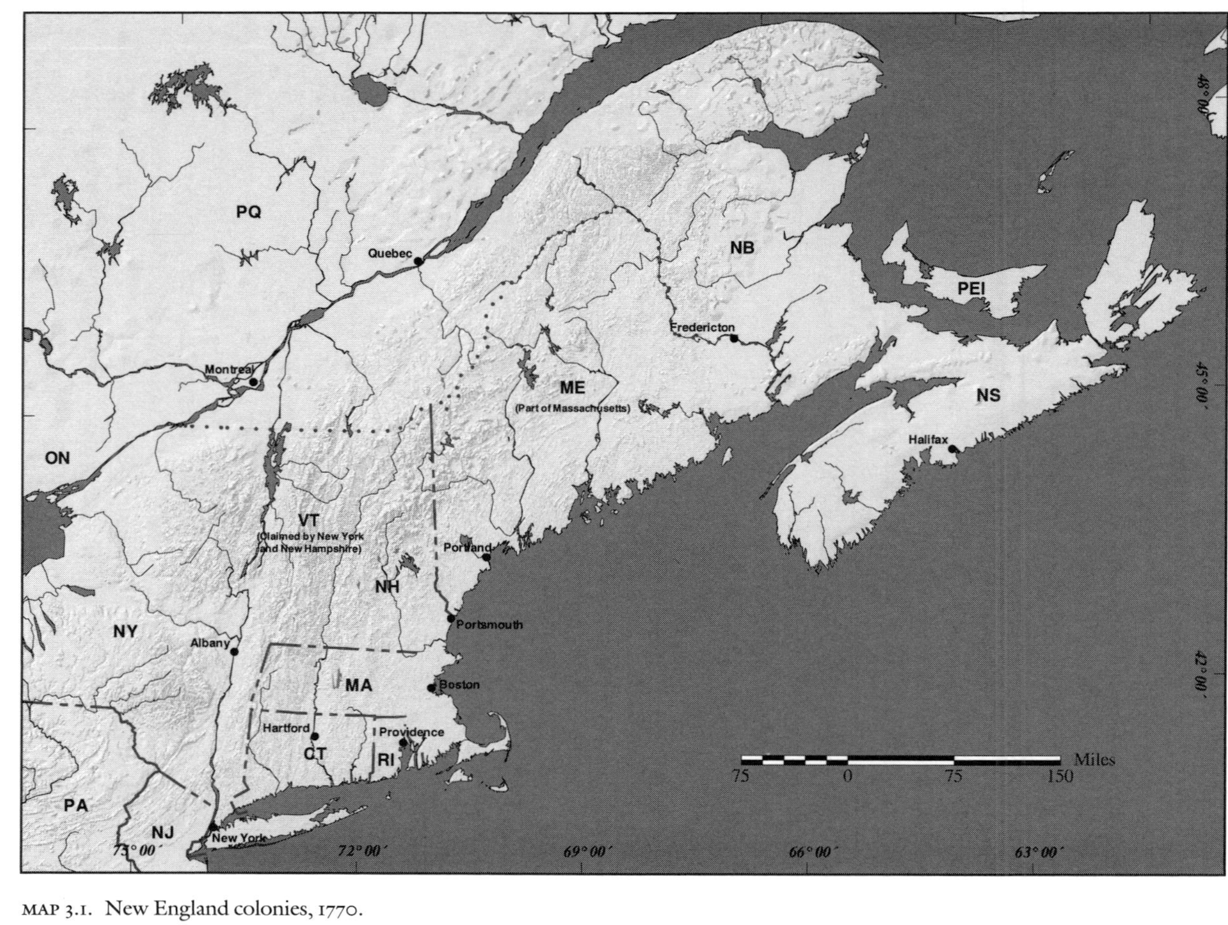

MAP 3.1. New England colonies, 1770.

New York promptly asserted its authority over the lands, based on the 1664 grant. The issue, like the boundary dispute before it, was sent to the king in 1753 for a final decision. The war postponed settlement of the dispute. It was not until July 1764 that the King in Council ruled in favor of New York. Thus, Vermont—then referred to as the New Hampshire Grants—was actually part of New York. Between the end of the war with France (1760) and the royal order, however, Governor Wentworth had been busy. He created 112 new towns in the Grants between 1760 and 1764, though both colonies had agreed to make no new claims until the king had settled the matter. His combined grants totaled nearly three million acres, approximately half of Vermont. The grants were mainly in the Champlain Valley and Connecticut River Valley, with the more rugged Green Mountains and its foothills still ungranted. Although half of Vermont had been granted, it was sparsely settled. In the mid-1760s, only forty to fifty thousand acres were actually held by settlers; the rest was held by speculators. In 1769, only 51 of the 128 townships had any residents at all.

The towns established in Vermont were done so under the New England proprietary system. A grant, usually of roughly 25,000 acres, was made to groups of proprietors, who owned the land together and jointly controlled its development and distribution. Much of the land was distributed to members of the group, though other land was granted to new members and those who came to live in the town. Once all of the land in the grant was distributed, which often took a number of generations, the propriety dissolved. The key to this approach was the condition of actual settlement and occupation of land; title was usually given—not sold—to proprietors under this condition of occupancy. The main activities of proprietors were the settlement of the town, division of the land, and control of the common field system. Settlers were attracted through grants of small tracts of land or small bounties. Proprietors usually first sought to attract millers and blacksmiths through incentives. Lots were also granted for religious and educational purposes.

By the time this proprietorship system developed in Vermont, the importance of occupancy had declined and the proprietors were involved more for speculative reasons. For example, in Windsor, a 1761 New Hampshire grant in the Connecticut River Valley, only three of the original fifty-nine proprietors ever settled in the town. By 1771, when the proprietary group dissolved, there were fewer than twenty-five proprietors. Each received 350 to 370 acres, including three large house lots totaling 130 acres near the center of town. Limited work had been done on roads and bridges in the town. This distribution of land, especially the house lots, had a very important implication for the nature of settlement in Windsor: It led to homes being scattered throughout the town, rather than being more com-

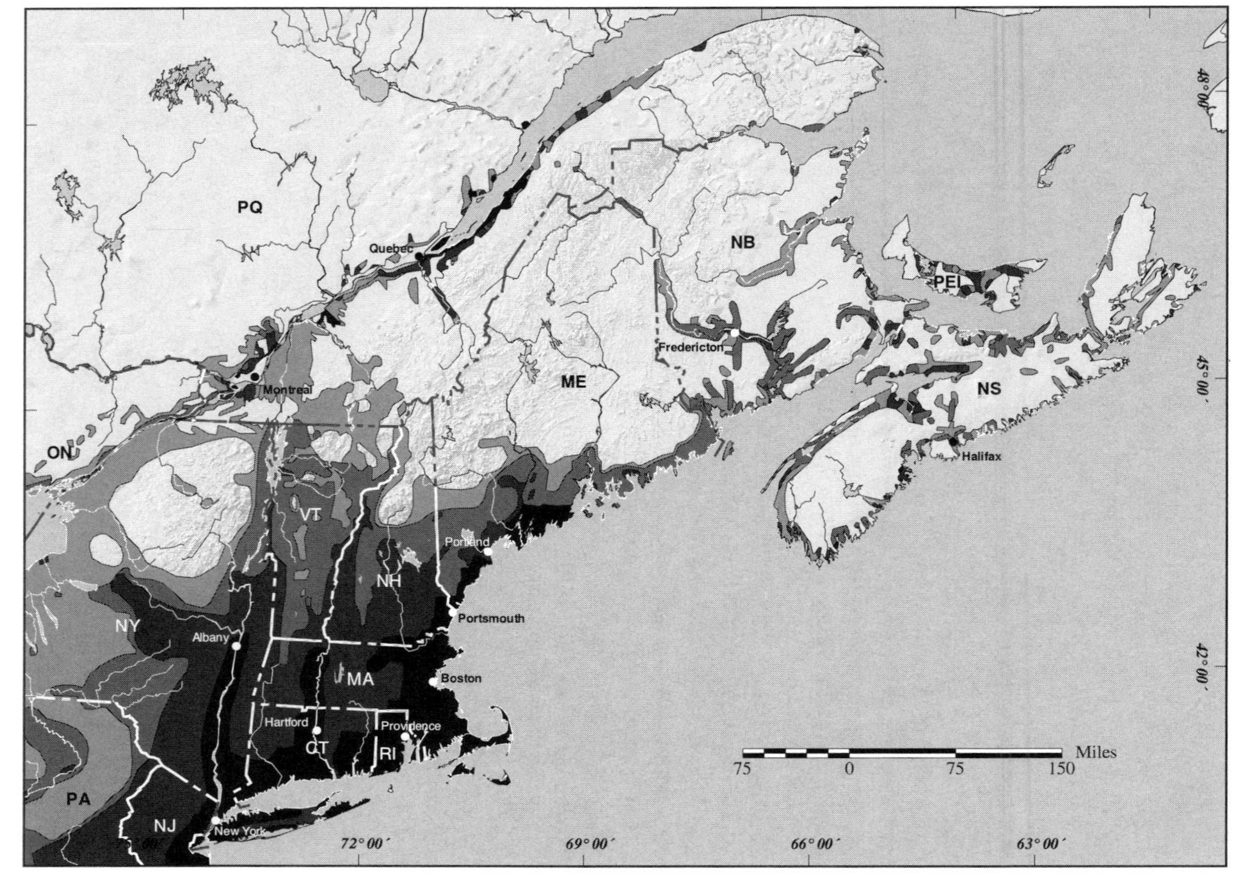

MAP 3.2. European settlement of the Greater Laurentian Region, through 1815. *Key:* ■ 1700; ■ 1775; ■ 1791; ■ 1815.

pact, as in the villages established earlier, when proprietors received small house lots in the village.

As these events indicate, land speculation played a key role in the early history of the state. The conflict between speculators and settlers who held New Hampshire claims and those who held New York claims was the major force leading to the creation of Vermont. Landholders who held titles from New Hampshire, most famously the Allen brothers (Ethan, Heman, Ira, and Zimri), led the fight against New York rule. The Allen brothers, like many other speculators, bought New Hampshire grants very cheaply following the king's ruling in 1764. At one point, they held nearly eighty thousand acres in the Winooski River Valley. If New York achieved authority over these lands, they would lose their entire investment. If they could prove the New Hampshire grants to be valid, their profits would be substantial.

Between 1763, when it was estimated that three hundred people lived in the region, and 1775, 12,000 people settled in Vermont and four thousand more were born there (see map 3.2). Although northern New England was generally growing much faster than southern New England at this time, the northern region, due to its natural characteristics, would have been settled by Europeans much later, even without the warfare. This land was colder, harsher, and more mountainous; was not as well suited to agriculture; and was removed from the sea, a crucial source of food and transport. Those moving to the region consisted primarily of cohesive groups heading north to settle together. These settlers took one of four main routes into Vermont: up the Connecticut River; up the Deerfield River into the Green Mountains, through the mountains, and into southwestern Vermont; following the Housatonic River from western Connecticut into the Berkshires and then into southwestern Vermont through the Valley of Vermont; and the Crown Point Military Road. Once these settlers arrived, they proved to be quite persistent. Of those in Vermont in 1775, by 1791 70 percent still lived in the same town and 95 percent remained in the state. Land was also widely owned in Vermont. Between 1773 and 1789, 80 percent of adult male heads of households owned land. Few owned more than three hundred acres; one third owned less than one hundred acres. Much of the land was not cleared, though, so land ownership did not guarantee prosperity to its owners.

As the population in Vermont grew, mainly through settlers who moved from Connecticut and western Massachusetts, New York attempted to establish control over the territory. In 1766, New York created Cumberland County (current Windham and Windsor Counties); in 1770, Gloucester County, north of Cumberland; and in 1772, Charlotte County, which encompassed Vermont and New York north of the Batten Kill and west of the Green Mountains. The southwestern corner of Vermont remained in Albany County. By 1777, New York grants in Vermont totaled 2.4 million acres. Those living to the east of the Green Mountains acquiesced to New York's authority. But to the west of the mountains, where most of the large speculators held their lands, resistance was significant from the outset, especially when New York began regranting some of the lands already granted by New Hampshire. A revolution of sorts, centered west of the mountains, began against New York. Riots at courts administering New York's authority began in the early 1770s, culminating in the death of two men at the Westminster court in March 1775 (the Westminster Massacre). An interesting footnote is that the first riots were related to the prosecution of individuals charged with destroying protected mast timber. These events led Cumberland County to renounce New York's authority over Vermont and seek to be "either annexed to some other government or erected and incorporated into a new one." This was the first public mention of a new state.

New York sought the aid of the British Army against the Green Mountain Boys in 1773, but the king and army dismissed the request, fearing that such a show of force would undermine civil authority. New York did little to secure its authority over the land, even though the riots never involved more than 150 rebels. Prior to the American Revolution, many in the Grants believed that the British were about to establish a new colony there, but nothing came of this. Once the Revolution began, New York and Vermont concentrated on the British foe, though New York did request that the Continental Congress recognize its jurisdiction over Vermont. In May 1776, those advocating an independent Vermont petitioned the Continental Congress to recognize the New Hampshire Grants as separate from New York, but this petition was withdrawn when it became clear that it would be rejected. The Continental Congress did not rule in favor of New York, though, and this issue—"the Vermont problem"—remained on the agenda of the federal government through 1791.

The Founding of Vermont and the Vermont Republic

By 1777, the movement for independence was supported on both sides of the Green Mountains. In January, a meeting of towns drafted a declaration of independence from New York, proclaiming the creation of New Connecticut. This proclamation did not meet with unanimous support. Some in the Grants continued to support New York's jurisdiction; others, especially in the Connecticut River Valley, hoped that New Hampshire would claim authority over the towns on the west bank of the river. By April, however, New York had adopted a new conservative constitution, leading many New York supporters in the Grants to now embrace the idea of a new state, while New Hampshire made no moves to further its authority. In June, a larger meeting of Vermont towns proposed a Declaration of Rights and Frame of Government. At this meeting, Vermont was adopted as the name for the new state (as suggested by Dr. Thomas Young of Pennsylvania, a friend of the Allens) since a group in northeastern Pennsylvania seeking to form a new state there was already calling itself New Connecticut. In July, this constitutional convention reconvened, and the new constitution was accepted unanimously. This constitution was interesting in a number of respects: It made Vermont the first state to institute universal manhood suffrage and the first to outlaw slavery. In terms of land, it put the community interest ahead of individual interest—declaring "that private Property ought to be subservient to public uses." But most important, Vermont had declared itself independent of England, New Hampshire, and New York.

Creating a new state or republic was not an easy task. Vermont had been without a well-respected and well-functioning government since its inception. Unlike the other states, Vermont could not rely on past colonial charters or a preexisting political community. So it had to convince its inhabitants that it could be a legitimate government. Since Vermont remained outside of the United States until 1791, it was also unique among the states in that it was created by a free people based upon their consent. As historian Peter Onuf writes, "Vermont was the only true American republic, for it alone had truly created itself."

The following March, the General Assembly met in Windsor and Vermont began its fourteen-year existence as an independent republic, though from the outset, Vermont wanted to join the United States. At first this was impossible. Among its neighbors, Massachusetts did not oppose statehood, but New Hampshire still hoped to gain some of Vermont, and New York was bitterly opposed since it still claimed all of Vermont. This "Vermont problem" was a significant one for the Continental Congress and later congresses. The national government did not want to alienate New York, an important state in the union. Furthermore, it did not want to encourage new states being carved out of existing ones. Indeed, article 4, section 3 of the United States Constitution declares that no new state can be created out of an existing state without the consent of the existing state and Congress. On the other hand, members of Congress supported democracy and the rights of people to organize their own government (as the United States had fought the British to do), and they wanted Vermont to join the union rather than become a British colony.

During its independence, Vermont explored two options to reduce its vulnerability while its hopes for statehood were stalled. Reuniting with England as a colony was the first. This alternative was most fully pursued in the Haldimand negotiations in the early 1780s. By doing so, Vermont political leaders were covering their bases and putting pressure on the United

States to add Vermont to the union. The British loss at Yorktown and the subsequent Treaty of Paris led to the collapse of these negotiations, but this was not the end of the British option. As late as 1788, Ethan Allen was still exploring Vermont's return to the British Empire with the governor of Quebec.

A second approach Vermont explored—and did act upon—was to extend its territory. A larger republic, the argument went, had a better chance of survival than the existing Vermont. From 1778 through 1782, Vermont acted three times to extend its territories, twice to the east and once to the west.

In March 1778, sixteen New Hampshire towns along the Connecticut River petitioned Vermont to become part of the new republic. This was approved by popular vote in the towns of Vermont, and in June 1778 the General Assembly voted to admit the towns, creating the first Eastern Union. This union came under fire almost immediately, from both outside and inside Vermont. New Hampshire asked Vermont to dissolve the union, an action also favored by political leaders from west of the Green Mountains because the union reduced their power. The Eastern Union was essentially killed in October 1778 by a vote in the General Assembly, partially in hope that this would lead to statehood. Not until February 1779, though, did the legislature vote to officially end the union. In the interim, a number of New Hampshire and Vermont towns in the Connecticut River Valley met to explore the possibility of creating a new state centered on the Connecticut River, consciously acknowledging the greater importance in their daily lives of the boundaries of the watershed than the existing political boundaries. By March, these towns came to favor New Hampshire taking complete control over Vermont, creating a state centered on the Connecticut River Valley. New Hampshire never warmed to the plan, and the issue simmered until the second Eastern Union was established.

In April 1781, a second Eastern Union was established with thirty-five towns along the Connecticut River. Vermont created Washington County—entirely in southwestern New Hampshire—and enlarged Orange and Windsor Counties into New Hampshire to administer this new territory. Two months later, the Western Union with twelve New York towns was created. Bennington and Rutland Counties were extended into New York to administer this territory. Ethan and Ira Allen envisioned an even vaster Greater Vermont: To the west, the border would follow the Hudson River, then north through the Adirondacks to Canada; to the east, in southern New Hampshire, the towns along the Connecticut River would be part of Vermont, and in northern New Hampshire the new border would extend all the way to the border of Maine (see map 3.3). Indeed, Ethan Allen even talked of annexing Berkshire County from Massachusetts. Obviously, these

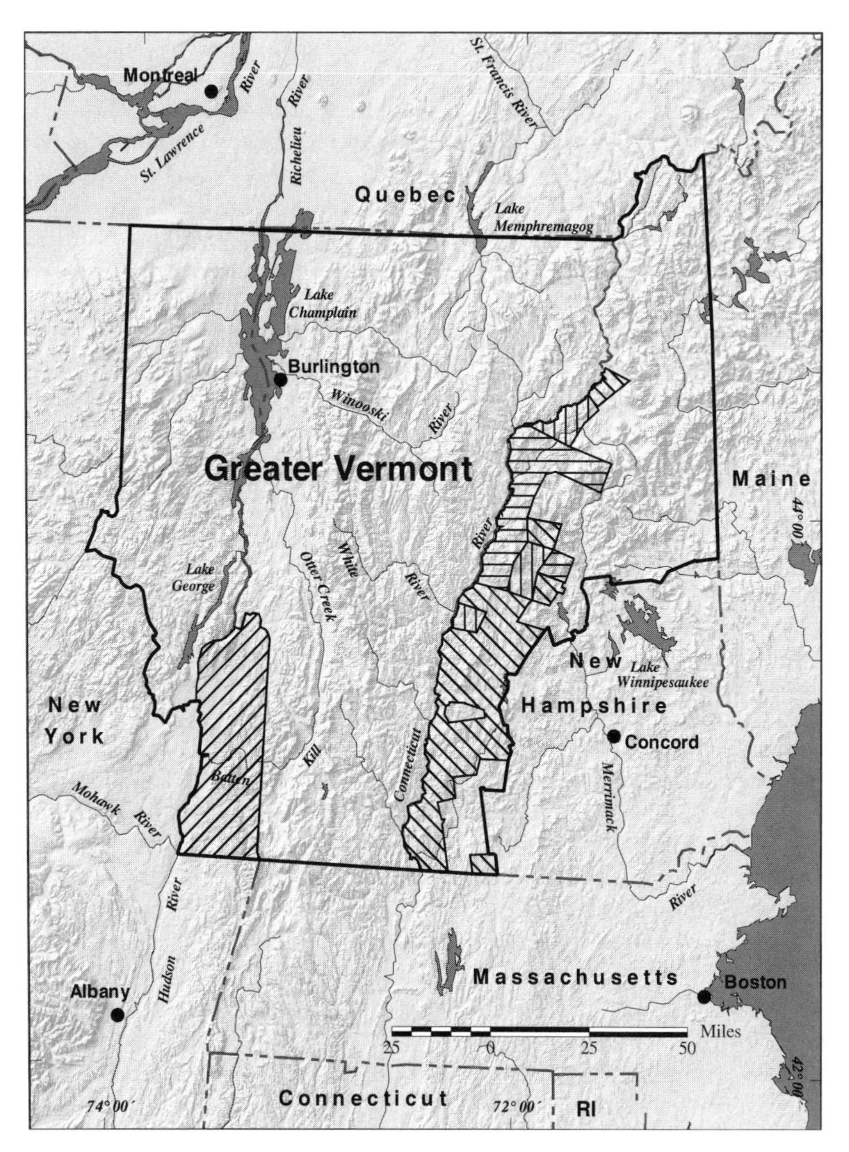

MAP 3.3. Greater Vermont Republic, 1778–1782. *Key:* First Eastern Union; Second Eastern Union; towns in First and Second Eastern Unions; Western Union.

unions were strongly opposed by New Hampshire and New York, but since Vermont was an independent republic, the neighboring states could not appeal to some central government to bring Vermont into line. Nevertheless, the Eastern and Western Unions, part of nascent Greater Vermont, did not last even a year. George Washington wrote to Vermont, asking it to dissolve the unions. If it did so, statehood was virtually assured. Largely in response to this request, as well as in light of the ends of the Revolution and the British option, the Eastern and Western Unions were dissolved in February 1782. Statehood would have to wait another nine years, though.

"The Vermont problem" languished due to New York's claims to the state, the relationship of the Vermont question to the future disposal of the western territories, and the difficulty of any new state's being admitted to the union under the Articles of Confederation. With the implementation of the Constitution in 1789, which made the admission of new states more viable, and the continued growth of Vermont, the stage was set for Vermont to finally join the union.

Before turning to statehood, a brief examination of the effect of the Revolution on the Vermont landscape and independent Vermont's land policy is in order. The only battle in the Revolutionary War in Vermont was at Hubbardton; the Battle of Bennington in 1777 was actually fought just across the state line in New York. There were numerous raids throughout Vermont by Indians allied with the British, resulting in most settlers abandoning the northern half of the state. Middlebury and Royalton were burned, and farms along Lake Champlain were destroyed. Nonetheless, the Revolution had a very limited effect on the Vermont landscape. After the war, the Continental Congress passed a resolution recommending that the Continental Army invade Vermont, but this resolution was rejected by General Washington in February 1783.

During this period of independence, Vermont implemented its own land policy to dispose of lands within its borders. It recognized the charters to land granted earlier by Massachusetts, New Hampshire, and New York (five Vermont towns were originally chartered by New York), but this still left substantial lands within the state ungranted. Vermont sought to generate revenue from the sale of these lands—plus lands confiscated from Tories—in order to finance the government. The new town grants, 128 between 1777 and 1793, were sold to proprietors. For instance, Hyde Park was sold to a group of sixty-five proprietors, with five public shares reserved (two to support churches, two to support local education, and one to support a state university). As in the past, proprietors were required to settle on and cultivate the land. In addition, "all pine timber suitable for a navy should be reserved for the use and benefit of the freemen of the state," a provision copied from other New England charters. Despite the occupancy requirement, only two of the original proprietors were ever known to have lived in Hyde Park. By 1813, when the proprietary group dissolved, each right was entitled to 340 acres. As was the case in Windsor earlier, the distribution of land did not include small house lots in a central village, leading to scattered homesteads. A further difference from the older proprietary system was the lack of a common field system or common pasture. The Hyde Park propriety made no effort to encourage mills in the town and did very little work on roads. By the early 1800s, the cooperative-proprietorship approach to land settlement, one that never really functioned fully in Vermont, was dead. Overall, this approach to land distribution meant that Vermont towns had very different characteristics than the towns in southern New England; chiefly, they had no common fields and lacked central villages with small house lots. Yet in Vermont as elsewhere in New England, the number of large estates and large landholders was quite small in comparison to other places in the United States, and especially in comparison to England. There was virtually one uniform agricultural class, made up of freeholders owning their own land scattered across the landscape.

Vermont Joins the United States

In October 1790, a settlement that cleared the way for Vermont statehood was finally reached. Vermont would pay New York $30,000 to compensate New Yorkers who held title to land in Vermont, and New York would give its consent for Vermont statehood. On 4 March 1791, Vermont became the first new state admitted to the United States. At statehood, Vermont had a population of 85,341, nearly triple its population in 1781.

Among the many things that resulted from Vermont's becoming a state, two are of special interest for our story. First, in 1793 Vermont amended its constitution, repealing article 21 of the 1786 Constitution, which granted all residents the right to "form a new state in vacant countries, or in such countries as they can purchase, whenever they think that thereby they can promote their own happiness." Hence, the right of citizens to do what Vermont did—create a new political entity within the borders of an existing political entity, without that existing entity's permission—became a right no more. No new states would be carved out of Vermont unless the governments of Vermont and the United States agreed. Second, a property tax of a penny an acre was established to raise the funds needed to compensate New York. This first property tax of Vermont's statehood was to pave the way to increased reliance on property taxes to fund government, especially local governments. Such taxes have had, and continue to have, significant effects on land use in Vermont.

As a new state, Vermont's eastern, western, and southern borders were now clearly established (though in 1880 there was a slight border adjustment in which New York gained a small piece of land from Rutland County). There was still some fluidity to the north, however. The Treaty of Paris in 1783 had reaffirmed the boundary between Canada and the United States at forty-five degrees north latitude, where Vermont and Quebec bordered, a border first mentioned in 1763 when Quebec became a British colony. The Treaty of Paris boundary was set thus even though Vermont was an independent republic at the time. Despite this relatively clear border, disputes and confusion over the Quebec-Vermont border persisted through 1842.

In the early 1790s, the confusion focused on Alburg, claimed by Vermont, New York, and the British. The British had garrisons at Point-au-Fer and at Dutchman's Point in North Hero, both on land claimed by Vermont. This dispute was settled by the Jay Treaty, which required that the British give up all forts in the United States by 1796. The Treaty did not settle the border problem, though. Vermont claimed that it had been set south of the forty-fifth parallel; the border should be further north, giving Vermont more land. The next major border problem arose in relation to the War of 1812. After this war, the border ceased to be a military problem, and the legal boundary between Vermont and Canada was finally and formally established by the Webster-Ashburton Treaty of 1842, which based the border on a line drawn between Quebec and New York in 1771 and 1772. The boundary is between 0.25 and 1.1 miles north of the forty-fifth parallel along Vermont, giving the state an extra sixty-plus square miles of territory. With this treaty, all of Vermont's current borders were now firmly in place.

From 1600 to 1800—especially after 1760—the most dramatic changes to the Vermont landscape were taking place since the glaciers receded more than 12,000 years earlier. The dominant human culture of the region, the Abenaki, was replaced by a European culture. The human population in the region increased over fifteenfold. A new political entity—Vermont—was born, first as an independent republic and later as one of the United States, to govern this territory. We have discussed some of the effects these changes had on the landscape of Vermont. In the next chapter, we will examine these changes in more detail: changes to the forests; changes driven by agriculture, industry, and mining; and changes related to more people, vaster market systems, and new methods of transportation.

4

A Landscape Transformed

The Vermont Landscape from Statehood through the Civil War

AS VERMONT SETTLED into statehood, the forces affecting the landscape accelerated. The human population grew, and these people cut forests, brought more and more land into agriculture, expanded mining and industry, and developed a transportation system of roads, canals, and railroads that both simplified travel in Vermont and also further tied the state to larger regional, national, and international trends. All of these changes can be viewed as part of the second ecological revolution to occur in New England since the arrival of the Europeans, the capitalist ecological revolution. During this revolution, which lasted from roughly 1800 to 1860 in Vermont and much of northern New England, production became oriented toward profit rather than subsistence, and nature became something to be mastered as a source of wealth. It was during this period that the most wholesale human alterations of the Vermont landscape were made. Afterward, patterns were reinforced and changes were of a more local nature, part of a larger dynamic mosaic of change and recovery. These subsequent changes, though, were nothing like those of the first wave of European settlement.

Changing the Landscape: Cutting the Forests

The forests of Vermont were cleared for four main, interrelated reasons: lumber, farmland, fuelwood, and potash. Although the land was unsettled by Europeans at the time, laws regulating the use of the forest in Vermont date back to 1639 in New Hampshire and 1665 in New York. The first laws dealt with when forests could be burned, with later laws covering the cutting of trees on land belonging to others, limiting the diameter of trees cut for firewood, and regulating the quality of timber products. The most important and widespread laws dealt with the use of forest products in the naval trade with Britain: white pine, pitch, rosin, tar, and turpentine. Since Vermont was barely colonized by 1776, it did not play a major role in this trade nor in the regulatory policy that hardened opposition to British rule, although several ship masts over ninety feet in length and over four feet in diameter were floated down the Connecticut River from above Bellows Falls. In addition, many loggers came south from Canada to cut trees in the Champlain Valley and float them north. Indeed, in the 1760s most of the oak staves exported from Quebec were from trees cut in Vermont.

For the most part during the early days of settlement and statehood, there was much more forest than there was demand for forest products. Most of the forest was viewed as an obstacle to farming rather than as a resource to be harvested, and these excess trees were usually burned. The lumber that was cut was used on homesteads to construct furniture, houses, fences, and barns, or was sold or bartered in nearby villages. Construction of a sawmill almost immediately followed settlement, and the clearing of land for farming kept these mills supplied with timber. It was not until after an area had been settled and its land cleared that a separate lumber industry developed in a region. The exception was the continuation of cutting and sending lumber north to Canada. In the early 1800s, with the British Navy cut off from its other main source in the Baltics due to war, this trade in white oak and white pine flourished. Over half of the oak and most of the white pine exported from Quebec to England at this time was cut in the Champlain Valley. This northern lumber trade came to an end with the War of 1812 and the opening of the Champlain Canal in 1822 (discussed below). By then, though, the best accessible stands for naval use had already been cut.

At different times during this period, Burlington was one of the great lumber centers in America. Logs were floated down rivers and gathered in the city and then rafted up Lake Champlain (which drains northward), through the Champlain Canal to the Hudson River, and on to New York. This phase of the lumber trade had nearly ended by 1840, as almost all the accessible sawtimber stands had been cut. Although thirty years later there were only six hundred mills—four hundred fewer than in 1840—the value of goods produced had increased tenfold, thanks in large part to railroads connecting Burlington to Boston and beyond. Much Canadian lumber was milled in Burlington and shipped on to markets throughout the northeast. Railroads also made once inaccessible Vermont forests commercially attractive, leading to another logging boom in the state. Indeed, Burlington was the third most important sawmilling center in the country in the 1870s, after Chicago and Albany. Furthermore, railroads stimulated numerous small-scale wood-products concerns throughout the state, which made things such as butter firkins, clothespins, ladders, and wooden bowls.

By the mid-nineteenth century, cutting of the forest was driven primarily by the market needs of a growing economy and nation, not subsistence farm needs. This led both to increased cutting and to more selective cutting, which in turn changed habitats and altered tree composition of the forests. For example, in Chittenden County, today's forest has more birch, hemlock, maple, oak, white pine, and other species than it had before intensive logging began; there are many fewer beech (which take longer to reestablish themselves) and spruce (favored by loggers). The timber industry that developed in northern New England established a pattern of migration for the forest-products industry nationally. Loggers would cut a region and then move on. From coastal New England, they moved to inland New England, including Vermont. When the best trees were cut there, the loggers moved on to New York and Pennsylvania, and from there to the Southeast and the Midwest, the Northwest, and now back to the Southeast (and to other countries as well). Directly related to this migration was the industry's disinterest in managing forests. It simply cut and moved on, often leaving a decimated forest behind. This nonmanagement approach helped stimulate the creation of the national forest system in the late 1800s.

This market logging exacerbated the reduction in forestland caused by clearing for farmland. In addition to obvious dramatic ecological changes in forests, the proliferation of sawmills and river driving of logs also disrupted the ecology of rivers (a process that continued in Maine into the 1970s) and of entire watersheds due to changes in water flows. By mid-century, logging and farming had fundamentally transformed the land. As environmental historian Carolyn Merchant writes, "The resulting environment was very different from that encountered by Pilgrims and Puritans. Thousands of acres of woodlands had disappeared, swamps had been drained, beaver ponds had vanished, streams and rivers had been dammed. Indians had been virtually annihilated by European diseases and warfare, and the forests of their ancestors were nearly depleted of the resources needed for subsistence." Yet to the great majority of the human inhabitants of New England, this new landscape was cause not for concern but for celebration. A land had been tamed and made productive. Nonetheless, the seed of conservation began to germinate in the soil of these ravaged forests. Localized wood shortages, flooding, and soil erosion worsened by the removal of forest cover, and problems with navigation led some, such as Vermonter George Perkins Marsh, to begin to focus on the negative consequences of these drastic changes. As he stressed in *Man and Nature*, published in 1864, humanity must start to conserve and utilize nature more wisely, or current human civilizations will suffer—perhaps collapse—as they have in the past.

The forests were of crucial importance as a source of fuel to settlers well into the nineteenth century. On a typical family farm, one to two acres per

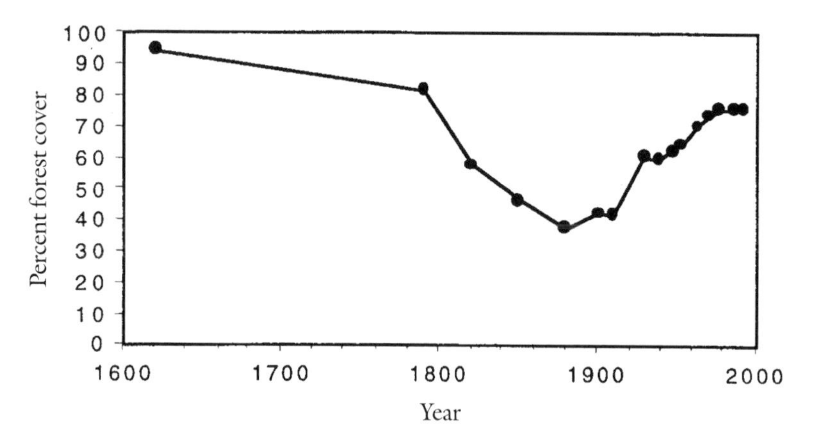

FIGURE 4.1. Estimated percentage of forest area in Vermont, 1620–1992.

year of woodland were required for fuel, with a forty-acre woodlot providing a continual supply. From a long-range perspective, fuelwood was the most important product taken from the forest, representing more than half the total volume cut as a commodity (more than double that of lumber). Wood was also converted to charcoal for various industrial purposes, especially iron making. The development of steam power further stoked the demand for fuelwood, and national per capita consumption peaked in 1840 at 4.5 cords per person (although total consumption in New England did not peak until three decades later). The demand for fuelwood to supply the railroads in the 1850s led to regional shortages. For instance, around 1850 Vermont's railroads were burning 63,000 cords of wood per year. This demand drove the fuelwood market until trains started to change over to coal, a transition that began around 1865 and lasted into the 1880s. The pressure on forests for fuel first began to slow when more efficient woodstoves started to replace fireplaces (ca. 1840), and again when industries and homes also began to switch to coal as their fuel source. This transition to coal—made for economic reasons, not for conservation—was slower in Vermont than elsewhere due to Vermont's more rural population, the difficulty in transporting the coal, and the still relatively abundant fuelwood.

Potash was made by burning hardwoods to a fine ash, leaching it with water to create lye, and boiling it down in large iron pots to a dense, ashy deposit. This was used in the manufacture of soap and glass, in the processing of wool and flax, and as a fertilizer. Always in demand and easy to transport, potash was one of Vermont's earliest and most important exports. There was, however, a major ecological problem with making potash: It removed organic matter—nutrients—from the ecosystem. (The importance of such nutrients in natural communities is discussed in chapter 7.) By 1840, the state led New England in potash production, though the market soon collapsed when an industrial process to make potash more cheaply was developed.

The decline in demand for fuelwood and potash, combined with the decline of the timber industry and farming, allowed the forests to begin to return. Ecologists and foresters have estimated that 95 percent or more of Vermont was covered by forest when the Europeans arrived. This forest area, they estimate, dropped to 82 percent by 1790, to 58 percent by 1820, to 47 percent by 1850, and bottomed out at 37 percent in 1880 before the percentage of land forested began to increase, though some argue that the maximum amount of cleared land could have been prior to 1850 during the sheep-farming boom (see fig. 4.1). It was in the 1870s that the Vermont landscape looked most different from how it looked in 1609; a state almost completely covered by forests had been substantially cleared of them.

Changing the Landscape: Agriculture

Vermont's soils are very rocky due to the glacial action described in chapters 1 and 2. In some places, the glacial till is hundreds of feet thick. The soil also tends to be acidic and low in nutrients and organic matter. In a national context, they are poor soils. Northern New England's soils can be divided into three general groups. The mountainous and hilly regions—the Green Mountains, Northeastern Highlands, Piedmont, and Taconic Mountains—have thin soil and are very rocky. By the 1830s, much of this land had been cleared for raising crops in small, irregular fields or for grazing livestock. The difficulty in working this land and its poor fertility were the chief reasons for its abandonment in the face of competition from the richer western lands. This land eventually returned to forest. The upland valleys in these regions have light and fine soil that warms up early in the spring and is good for farming. Last, the lower river valleys and the Champlain Valley are areas of smoother contour and lower elevation (below five hundred feet) and have proved better suited to mechanized farming. These areas remain the most important agricultural lands today and include some of the best in New England. Overall, roughly one sixth of Vermont has prime agricultural soils.

The climate in Vermont is nearly as inhospitable to agriculture as the soil and terrain are. Due to its inland location, Vermont is the driest state in New England, leading to occasional droughts that are problematic for farmers. In most of the state, only three months are frost free, and even in June, July, and August the crops are not necessarily safe. The summer of 1816 was especially cold: six inches of June snow in northern and eastern Vermont, numerous July and August frosts, and a hard freeze in September. Furthermore, winters are cold and snowy. Given the marginal nature of much of Vermont agriculture, relatively minor climatic shifts (e.g., the generally cooler temperatures of the late 1830s and early 1840s) have had significant effects on farmers.

There were three general periods of agriculture in New England in the 1800s, with the duration of these periods varying by location: (1) self-sufficiency, from settlement into the 1820s and 1830s (later for more isolated towns), when there was no real market for farm products and each household and town depended primarily on its own resources; (2) transition to commercial agriculture, from the 1820s through the 1860s, in response to the rise of manufacturing and demand for food and raw materials by the nonfarming population, especially in southern New England; and (3) decline, from roughly 1865 to 1900 (as late as the 1930s for isolated hill towns), due to increased competition from western farms.

Around 1800, Vermont farming was based on self-sufficiency, expedited by an abundance of relatively cheap land, and restrained by the high cost of labor. Little fertilizer was used on the land, livestock were not closely tended, and farm implements were still rough and clumsy. A typical settler might clear one to three acres of land per year. If things went well, at the end of four or five years, the farmstead would be well provided for in terms of food and shelter. This translated into about 50 percent of the land being cleared for crops, pasture, and structures; the remaining land was left in forest or was considered "waste"—wetlands, steep hillsides, or rock outcrops. During this period of establishing a farm, the settlers relied on hunting and gathering—berries, fish, game, nuts—to serve as key food sources. (Indeed, deer were hunted so much that the first state legislature established a closed deer season.) Since these new settlers did not know the land the way the natives did, their first few years were risky ones in terms of starvation and exposure.

All of the organic matter that had accumulated on the forest floor meant yields of wheat, corn, and rye were very good for the first few years of planting. Soon, however, the fields became exhausted from the constant planting. These fields were then abandoned and new land was cleared, or families moved to new places. Sometimes, grasses and red clover were planted as pasture on this exhausted cropland, but at first livestock was turned loose, and the animals gained most of their food from the forests and natural meadows. As defined by law circa 1810, certain possessions and goods were exempted from collection for payment of debts. These exemptions are a good indication of the basic level of subsistence at the time: one cow, one or two swine, six to ten sheep and their wool, a pair of oxen or a horse, hay to sustain livestock through the winter, all standing crops, twenty to thirty bushels of corn and grain, and ten to twelve cords of firewood. This subsistence could probably be achieved on forty-five to sixty-five acres. More established farms, obviously, included more animals, with a typical farm having fifteen cattle, fifteen pigs, ten to twenty sheep, and some poultry.

The chief foods produced on these subsistence farms were beef, pork, and mutton; butter and cheese; bread made from Indian cornmeal and rye; fruits (especially apples for cider) and vegetables (especially beans, squash, and turnips); and maple sugar and honey. During the early years, hunting, fishing, and gathering continued to supply a significant amount of food. Farmers had to trade for a few staples they could not produce, namely, salt, tea, coffee, molasses, and rum. Most of the clothing was homespun from wool and flax produced on the farm. The buildings and furnishings were almost all made by the family from the timber cut to clear the land and later from the farm woodlot. Small amounts of hardware, glass, and utensils were purchased. Last, fuel was supplied almost exclusively by firewood from the farm. The farms sold and bartered some livestock, wool, butter, and cheese to gain a small income needed for taxes and other limited purchases. During this self-sufficiency period, wealth was fairly equally distributed; since land was cheap, there was no class of wage earners. Farm labor was supplied almost exclusively by family members.

The transition to commercial agriculture was fueled by many changes. Improvements were made in transportation, education and information, and tools and machines. Cities and western land successfully attracted large numbers of rural people. Although western farmers offered lower prices for such staples as beef, pork, wheat, and wool, the growth of population and industry in the Northeast led to a demand for perishable and bulk goods (e.g., dairy and hay) that favored New England producers over western ones. Also causing changes were land-inheritance practices. As farmers subdivided farms to pass them on to their sons, the farms became smaller and smaller, and eventually families could no longer support themselves. This problem of increasing populations and decreasing farm sizes could be dealt with through a more intensive, market-oriented farming or by settling new land.

All of these changes were related to the capitalist ecological revolution, and Vermont farming began the transition from extensive farming geared to subsistence to intensive farming geared toward market. This transition led to greater ecological stresses on the land due to monocultures and the problems that accompanied them, the contingencies of the market, and soil exhaustion. As Carolyn Merchant writes, "Attempting to produce greater surpluses with traditional techniques, they [the farmers] were unable to make worn soils yield greater fruits. Ultimately, those farmers who remained in New England resolved these systemic problems by individual decisions to change farming techniques, to specialize production, to hire wage laborers, to keep quantitative records, and consequently to become capitalist farmers." As capitalist farmers, decisions on what to grow and how to manage the land were now based on forces far beyond the borders of one's farm; indeed, the decisions were primarily affected by trends beyond the borders of Vermont.

Most important for the Vermont transition was an improved transportation system that could bring Vermont goods to growing commercial populations in the south and the north (see below). From its colonial beginning, the chief outlet for the Champlain Valley was through Lake Champlain to the north and Montreal. With the opening of the Champlain Canal in 1822, that trade shifted to the southwest via the Hudson River and Troy. In the East, locks and canals on the Connecticut River made it a viable route as far north as White River Junction. There was even some overland trade to Boston; indeed, as early as 1806, 15,000 cattle from Vermont were driven there. More generally, Vermont farmers sent beef, flaxseed and linseed oils, furs, grain, live cattle and horses, maple sugar, pork, potash, dairy products, timber products, and wool. In return, they continued to receive iron, molasses, rum, salt, dry goods, and tea.

Despite this nascent waterborne and overland trade, it was the railroad that really led to the switch to commercial agriculture. At first, these railroad connections opened up new markets in southern New England. But soon the railroads connected New England with the Midwest, and Vermont farmers had a difficult time competing with the products flowing in from there. Eventually, wheat, pork, wool, hops, butter, and cheese—most everything but fluid milk—came more cheaply from the West.

Economic historian Percy Bidwell summarizes this fundamental change for those working the land in Vermont: "Farming became a more speculative business, for to the already existent risks of weather conditions was added the risk of price-fluctuations. Thereafter success in getting a living no longer depended on the unremitting efforts of the farm family, aided by Providence, but to a large extent also upon the unpredictable wants and labors of millions of persons in the industrial villages, and in the newer farms to the westward." Furthermore, the landscape of Vermont was becoming integrated into regional, national, and international markets. Although the fur trade had already tied the Vermont landscape to contingencies far beyond its borders, commercial agriculture was to greatly broaden the range of those contingencies and their effects on the people, plants, and wild animals of Vermont.

Among the keys to improving the efficiency of New England farms and allowing at least some farmers to remain competitive with the western lands was the development of agricultural machinery throughout the 1800s. Much of this machinery—such as mowers, plows, reapers, and threshers—enabled the substitution of animal labor for human labor. Better hand tools were also developed, allowing farmers to accomplish more in less time. In addition to improved machinery, significant changes in field techniques began in the 1820s, including better tillage, an increased use of manure to restore soil fertility, and a more constant use of tilled land by eliminating summer fallows.

On established Vermont farms in the 1820s, the standard crops were grass and corn and usually wheat. Prior to 1820, some wheat grown in Vermont was traded out of state, but this ended with the opening of the Erie Canal and new lands to the west and the arrival of a wheat pest, the grain worm or midge, in the 1820s, both causing a significant decline in wheat farming in Vermont. Oats, rye, and barley were of minor importance. By 1840, hay was the primary crop in New England, and most agricultural attention was focused on livestock rather than grain. Cattle, dairy, and (as discussed below) wool markets all developed rapidly in the 1830s. Upland areas were now frequently seeded to clover and grass. A rotation system had been adopted by many farmers: corn followed by potatoes or oats, then wheat or rye, and then grass before starting the rotation over again. Agriculture in Vermont had now entered a period where increased farm productivity came not from increasing the number of acres farmed but by increasing the investment of labor and capital on existing acreage.

At the beginning of the Civil War, wheat, corn, oats, rye, and barley were all still cultivated in Vermont, but none was produced at a level of national importance and only the minor grains barley and oats were produced at greater levels than in 1840. Potatoes had proved popular for a time, with Vermont producing over 8 percent of the national crop in 1840, but a blight in 1843 led to a diversification into other root crops such as beets, carrots, rutabagas, and turnips. In response to competition from the West, many Vermont farmers tried their hand at specialty crops. The most successful was hops in the eastern part of the state. Vermont grew over 8 percent of the nation's hops in 1850, and in that year and in 1860 was a distant second to New York in amount of hops grown. The Vermont hops crop soon began to wilt under western competition, though; production was in decline by 1870 and had essentially stopped by 1900.

As livestock became the focal point of Vermont agriculture, hay developed as the key crop, both for farm animals and to sell in New England cities. The number of cattle in the state remained roughly the same throughout the mid-nineteenth century. Beef was produced mainly as a by-product of raising steers for yoke and cows for dairy. From Vermont, the beef cattle were usually driven to the markets in southern New England cities, chiefly

Boston. By 1870, though, western competition had essentially ended the beef trade of northern New England. Dairy grew steadily, replacing sheep as the state's most important agricultural product by 1850, when Vermont led New England in butter and cheese production. The state produced roughly 8 percent of the cheese in the country in 1850 and 1860, the most per capita of any state. Butter production, about 4 percent of the national total, was also the most per capita. Since Vermont had only about 2 percent of the dairy cows in the country, this demonstrated Vermont's market advantages in the perishable-dairy field. Meanwhile, a shift was under way from oxen to horses as the chief source of animal labor on the farm. Oxen worked better in the woods and at breaking new turf; they were also cheaper. But since little new land was being cleared, horses became preferred because they were easier to work with and faster. Vermont Morgan horses were famous nationally for their endurance, hardiness, and strength. Other breeds were also well respected, and cavalry buyers nearly stripped the state of them for the Civil War. Pigs were never particularly important in New England; most farmers kept just enough to clean up refuse.

In 1850, 2.6 million acres of Vermont land were classified as improved, the second most acres per capita but only a bit over 2 percent of the national total. In 1860, 2.8 million acres were improved, still second most per capita but now less than 2 percent of the national total. So, at the onset of the Civil War, Vermont was among the most agricultural states in the union, but its small size meant that nationally its production was not of much significance. This meant that Vermont would not set market trends but rather be at their mercy, a pattern that continues to the present.

The most visible example of Vermont's integration into commercial agriculture was the veritable sheep mania that raged from the 1820s through the 1860s. The Merino sheep, prized for its wool-producing characteristics, was brought to the United States from Europe in the 1810s, just in time for domestic sheep raising to supply the fast-growing textile mills of southern New England. The Vermont climate and soil were good for sheep, and tariffs helped the price of wool grow quickly from 1827 to 1835, leading Vermont farmers to focus more and more attention on raising sheep. With the lessening of self-sufficiency and the need for cash income, some farmers turned almost exclusively to sheep. During the peak of sheep raising in the late 1830s and early 1840s, every town in Vermont, except for a few towns in the Green Mountains and the Northeastern Highlands, had more than one thousand sheep, some over five thousand, and a few towns in the Champlain Valley and Connecticut River Valley had over 10,000 (see table 4.1). Around 1840, there were nearly six sheep per person in Vermont—over 180 sheep per square mile—and the state raised more sheep in proportion to its population than any other. Addison County was the center of this boom.

TABLE 4.1.
Human and Animal Populations in Vermont, 1800–1870

Year	Human population	Density (sq. mile)	Sheep population	Cow population
1800	154,465	16.7	—	—
1810	217,895	23.5	—	—
1820	235,966	25.4	—	—
1830	280,652	30.3	700,000 [est. 1828]	—
1840	291,948	31.5	1,681,819	151,757
1850	314,120	33.9	1,014,122	146,128
1860	315,098	34.0	752,201	174,667
1870	330,551	35.6	580,347	180,285

With 373 sheep per square mile, it is reported that it "raised a greater number of sheep and produced more wool, in proportion to either territory or population, than any other county in the United States." Vermont was even more reliant on wool than the south was on cotton.

This sheep mania began to wane in the middle 1840s for two main reasons, both rooted beyond the borders of Vermont. First, the protective tariff on wool was lowered in 1841, then eliminated in 1846, contributing to a decline in wool prices. And second, a gradual increase in wool production from the West led to a greater supply, further reducing prices. As the price of wool fell to half its mid-1830s level, many sheep farmers in northern New England began to reduce or even dispose of their flocks. There was a brief resurgence in the wool industry during the Civil War due to the demand for uniforms and the lack of southern cotton, but with the end of the war the Vermont wool era ended for good. Demand for wool declined, southern cotton was again available, and western wool was now joined by cheap wool from Australia and South America. By 1870, the sheep population was almost only a third of its peak of thirty years earlier.

The decline in the number of sheep was not matched by the decline in wool production. Improved breeding led to a 143 percent increase in the amount of wool per sheep from 1840 to 1870. This success made Vermont a center for sheep breeding; its Merinos were in demand all over the world. Sheep breeding, as well as Morgan-horse breeding (important from the early 1800s), remained an important part of Vermont agriculture into the 1870s.

One result of this overall decline in the sheep industry was the return of dairy. The prices for butter and cheese were much more stable than that of wool, and the arrival of railroads greatly improved the market for Vermont dairy products. By 1854, iced butter cars were running from Franklin County to Boston. As the urban dairy markets developed, farmers improved dairy productivity through increased supplemental feeds, especially root crops. The story of Vermont's dairy production will continue in the following chapters.

Agriculture was the predominant cause of the dramatic changes to the Vermont landscape that occurred from 1760 through 1870. Abenaki agriculture had had very little effect on the land, but as of 1861, 4.5 million acres of Vermont—roughly three quarters of the state—were classified farmland, and 3 million of these acres were improved. The forests were cleared, soils disturbed, and habitats destroyed. At first, this agriculture was small-scale. People came to Vermont to carve out homesteads and engage in extensive, largely self-sufficient agriculture. By 1870, agriculture had become integrated into national and international markets, and farmers were now vulnerable to market forces and government policies as well as climatic fluctuations. Increasing amounts of land were cleared for sheep, but the hard grazing of these flocks led to the final decline of some hill farm pastures. Indeed, the collapse of Vermont sheep raising due to outside forces was good for the land, since at its peak Vermont probably had more sheep than it could sustain and the land might not have later recovered quite so quickly had sheep continued to graze it. This integration into larger national trends also played a major role in the changing human population in Vermont, a story to which we now turn.

Populating Vermont: People Come, People Go

The population of Vermont grew 225 percent from 1790 to 1830. Over the next forty years, it grew only 17 percent (see table 4.1). From 1830 through the middle 1900s, commentators often reported that Vermont and its population were in decline. This is a misconception. Rather, there was a period of rapid population growth and then a long period of smaller increases alternating with periods of no growth. This differed from the steady growth in other places, but it was clearly not decline. Indeed, Vermont had the potential to have a relatively stable human population, but this potential was diminished by the state's increasing market connections.

During the earliest period of Vermont settlement, the only significant groups of people to leave the state were the perhaps five thousand Loyalists who left during and after the Revolution and several hundred "Yorkers" forced out for their support of New York's claims to Vermont. Substantial migrations took place within the state, though. People moved from the settled southern portions of the state to the open land of the north, and from the hills to the valleys. The uplands were settled first for a variety of reasons: The hills had fewer mosquitoes and illnesses than the valleys; the upland forest had less underbrush and was easier to clear; and the higher ground was drier and better for cultivation. But the valleys had streams and rivers, which meant mill sites and, as was later discovered, better soil. Mills for many purposes—grinding grain, sawing lumber, pressing cider, tanning leather, carding wool—were built at virtually every possible site. Hardships remained, though. Disease was still a major problem, though not anywhere close to the level that had afflicted the Abenaki. Weather-related crop failures also led to real hardships.

As the population continued to climb, it put a strain on the land and resources of the state. This was indicated in land values, which increased by 170 percent from 1791 to 1806. Furthermore, the lack of significant industrial development in the state made migration a logical option. At first, the little outward migration occurring was from the oldest settled counties of southern Vermont, where young people with no land headed off in search of places that were still undeveloped. In the early 1800s, the population of the state was young—over 50 percent were less than sixteen years old.

Around 1810, immigration to Vermont from the rest of New England came to a virtual stop and emigration out of Vermont significantly increased because of four main factors, the first of which was war. As the United States headed to war with England, a series of embargos greatly hurt Vermont's trade with Canada. Although much smuggling took place, the state's economy was harmed significantly. The war itself caused many to evacuate northern Vermont, and large numbers of these people never returned. Once again, as a border state, Vermont was especially sensitive to foreign-policy decisions. In response to the war, the New England states convened the Hartford Convention in 1814 to try to mold a regional policy. Many observers viewed the failed convention as a first step toward secession of the New England states from the United States. Vermont, however, played a limited role in this process, sending only an observer to Hartford.

Disease was a second force altering population trends. Dysentery, influenza, measles, pneumonia, scarlet fever, smallpox, and typhus were common, often with outbreaks killing significant numbers of people; for instance, more than forty people were killed by dysentery over the summer of 1803 in Stowe. An outbreak of spotted fever killed over six thousand, and tuberculosis accounted for one fourth to one third of the deaths in Vermont each year.

Third, the floods of 1811, which washed away roughly two thirds of the mills in Rutland and Windsor Counties, and the cold season of 1816 led others to try their luck elsewhere. And finally, the land in western New York and Ohio was a growing attraction. Vermont was no longer the frontier; the West was.

In the 1820s, Vermont's natural resources began to show signs of exhaustion due to the massive wave of human population and settlement that had crashed over the state since the late 1760s. Fish and game—important dietary supplements—became scarce (see chapters 7 through 9). Forests with trees valuable for lumber disappeared. Water power became unreliable with the decline of forest cover; floods were more frequent, and summer flows were less dependable. The soil started to wear out without fertilization. "The people," wrote historian Lewis Stilwell, "had mined the state rather than cultivated it." The sheep boom of the 1820s and 1830s also contributed to the decline in population in many towns as successful farmers bought out their neighbors to create larger sheep runs, while the neighbors headed west in search of new land. This emigration was still mostly from southern Vermont, where the land had been settled and the resources exploited the longest. Any growth in the southern and central regions was based in towns with mills, industry, or, eventually, railroad connections.

Vermont's population growth had slowed to 7 percent by the 1840s, the slowest growth rate in the union. This minimal growth came mostly from the arrival of French-Canadians and Irish to work in low-wage jobs. The state was isolated and had poor transportation, so it was bypassed by much of the industrial growth taking place elsewhere in the Northeast. Slow growth was not all bad, though. Indeed, the situation was quite good for those already in Vermont and working—the state produced more wheat, corn, and potatoes per acre, paid higher wages, and generated more wealth per capita than western states such as Illinois, Michigan, and Ohio. Yet for those not established, prospects in Vermont were bleak, and the West grew ever more attractive.

Population growth came to a virtual halt in the 1850s. Vermont grew by only 978 people (0.3 percent), and 54 percent of Vermont towns lost population. By a decade later, 60 percent had lost population. Most of these were upland towns in which the decline lasted well into the twentieth century. In 1860, of those born in Vermont still living in the United States, 42 percent lived outside the state (the largest percentage of any state). The state's rural population for the nineteenth century peaked in 1850, although its total population continued to grow, concentrated in a small number of cities and towns. As sheep farming gave way to dairy farming, more prosperous farms flourished, but the small farmer continued to decline. The coming of the railroad (see below) had a mixed effect on Vermont population. Towns along the lines grew; those removed from the railroads shrunk further. This especially favored valley towns and hurt hill towns.

Historian Harold Wilson focused on three reasons to explain this stagnating population growth and widespread emigration: the West, the growth of the city, and the upheaval caused by the Civil War. The West drew off the excess farm sons who could not find or afford land in Vermont. It also provided competition for Vermont agricultural products, competition that only grew with the railroads. Others moved to southern New England to work in the new factories there. For example, by 1835, 200 of Barnard's 1,800 residents had moved to industrial Massachusetts. This industrial migration was led by women, who were attracted earlier and in greater numbers than men to the cities. Vermont sent roughly one ninth of its population to fight in the Civil War (34,328 soldiers), causing a labor shortage at home. More than half of these soldiers did not return: Over five thousand of them were killed, and many others settled elsewhere after the war. Those who did return had a much broader worldview than they had when they left.

Other factors often cited to explain Vermont population trends in the 1800s include: (1) the exploitation rather than the sustainable use of the land —that is, the original economic gains came from improving the land, but farming itself proved very difficult for most to make a living from; (2) the difficult climate, soil, and topography; (3) the transition to pasture agriculture, which cannot support a large population; (4) the high birthrate of the early years, which led to a growing population with high-priced land and no other prospects, making emigration the only option for many; (5) the hard character of farm life; (6) the broadening of horizons through education; and (7) the various social and religious movements of the time.

Although Vermont was a rural, agricultural state throughout the first seventy years of the nineteenth century, certain towns and cities became centers of commerce, industry, and population. In 1800 and 1810, the largest towns —all with populations between two and three thousand people—were located in the Connecticut River Valley (Guilford, Springfield, Windsor, and Woodstock) or in the Valley of Vermont (Bennington and Rutland). By 1820, Middlebury was the largest town (at just over three thousand residents), followed by Burlington. Middlebury remained the state's most populous city in 1830, but from 1840 on Burlington and Rutland became the two largest cities of the state. Other cities surfaced in the top five through 1870 as well: Brattleboro, Montpelier (designated the capital in 1805), Northfield, and Saint Albans. Burlington's growth was fueled greatly by the completion of the Champlain Canal, which helped the city become a major inland port for trade to the north and south. Burlington had several industries too, with mills along the Winooski River processing grain, gypsum, timber, and textiles. Burlington was still no metropolis, though, with fewer than 14,000 people in 1870.

In examining county populations, from 1800 to 1840 the four most populous counties were the same: Rutland and the three Connecticut River Valley counties of Orange, Windham, and Windsor, with Windsor always the most populous. In 1850, Chittenden County cracked the top four and by 1870 was second most populous, behind Rutland. By that time, population had shifted northward, with Washington County in the top four in 1860

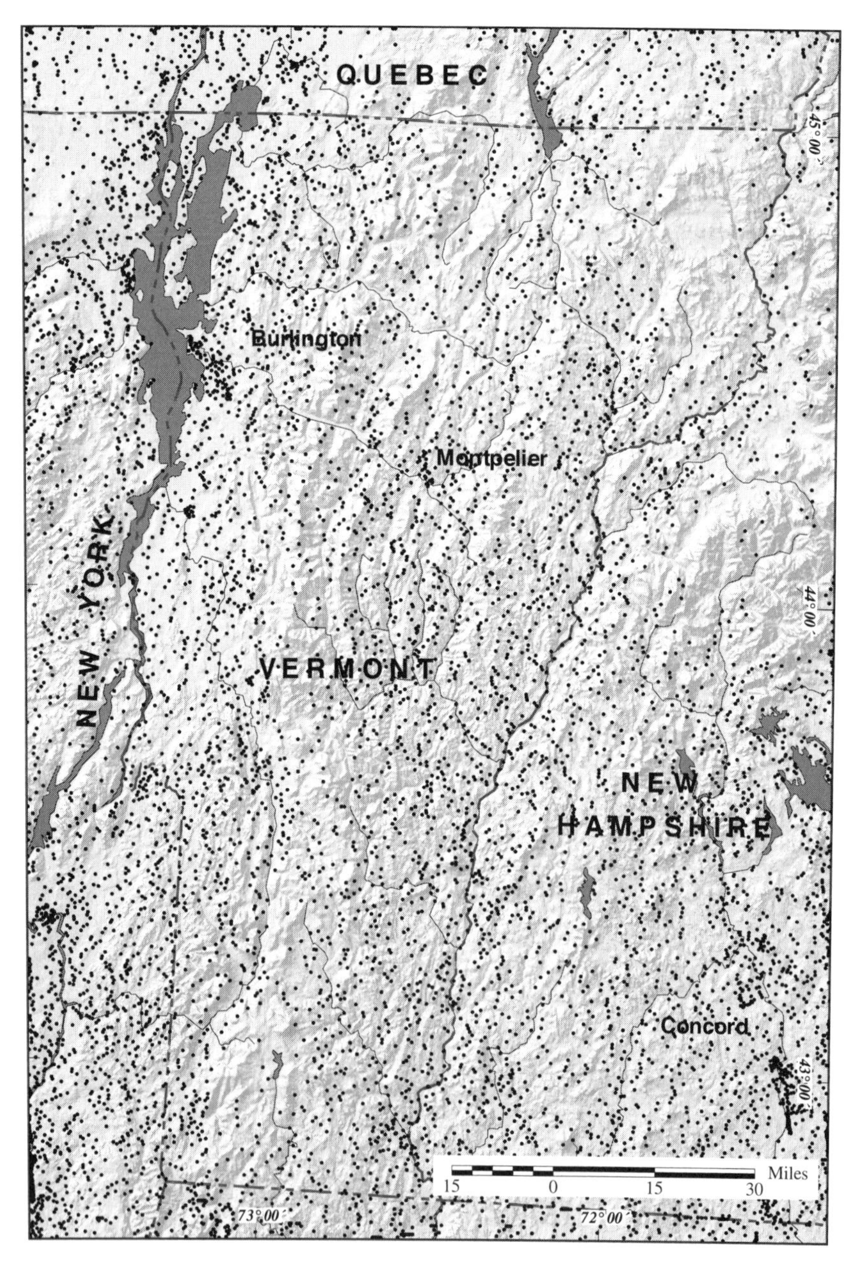

MAP 4.1. Greater Vermont population density, 1850. Each dot (•) represents one hundred people.

and Franklin County fourth largest in 1870. Vermont's population was greater than Maine's only in 1800, yet Vermont's population density always exceeded Maine's. Vermont exceeded New Hampshire in population in 1810, 1830, and 1840 and exceeded it in population density in those years as well, though from 1810 to 1860 the two states' population densities were quite similar (see map 4.1).

To summarize, Vermont was one of the fastest growing states in the country from 1790 to 1830. It was the northern frontier, and immigrants were attracted by the open land suitable for subsistence agriculture—the dominant way of life in New England at the time. That a native Vermonter was not elected governor until 1835 was symbolic of this mobility. There was also mobility within the state, as people headed north to open lands, to the valleys for mill sites and railroad access, and to the emerging cities. Burlington, the center of Vermont's only real urban area, became the state's most populous city in 1840. The state's overall population growth, however, came to a virtual stop in the 1850s. Vermont had become too populated for subsistence farming, and it was a poor competitor in commercial agriculture and manufacturing. Emigration from the state increased, and the landscape of Vermont—which had supported native peoples for thousands of years—showed signs of being worn out less than one hundred years after the surge of European habitation had begun.

Mining and Manufacturing

One does not usually think of Vermont as a major mining state, yet mining has played an important role in the state since the late 1700s. The continental collisions more than 500 million years ago (see chapter 1) created deposits of minerals in Vermont, as such geologic forces have throughout the world. The earliest important mineral in Vermont was iron, with the first forges built in the 1780s. The iron-ore beds were scattered throughout the state, though most of the major ones were in the Champlain Valley on the flanks of the Green Mountains and the eastern Taconic Mountains: Brandon-Pittsford-Rutland-Tinmouth, Bristol-Monkton, Dorset-Manchester, and Highgate-Swanton. The ore mined was of two types: bog ore, created in thin beds by erosion and deposition, mined from shallow, marshy areas; and upland hard-rock deposits. Prior to 1800, these ores served the needs of settlers on the frontier, being used to produce hardware, horseshoes, and the like. As Vermont became integrated into broader markets in the early 1800s, the demand for Vermont iron increased and Vermont became a major iron producer—indeed, the Monkton Iron Company in Vergennes claimed to be the largest ironworks in the United States circa 1813 (due to a contract to make cannon shot for the War of 1812). Other products made during this boom period included cast-iron stoves, machine gears, plows, and potash kettles.

The iron industry was in decline by the 1830s and more or less at an end by the 1860s. The main reason for this decline was that, as in other industries, the market undid the Vermont iron industry, which could not compete with producers located elsewhere. This was because of the lower quality of the Vermont iron ores, the susceptibility of mountain-stream mill sites to frequent flash floods, the decline of forest supplies of charcoal, the state's distance from major markets, the long winters, and the disappearance of certain markets (e.g., the one for potash kettles).

Four other major mining ventures were under way by 1815, three of which are still active today. In 1793, copper ore was discovered in Strafford, though mining did not start until 1821 in nearby Vershire. Mines were developed there, in Corinth and in Strafford and were a major—perhaps *the* major—source of copper in the United States until deposits near Lake Superior were developed in the 1840s. Even thereafter, the Vermont mines remained important suppliers, producing 1.5 million pounds of copper in 1870 (compared to 4.5 million pounds from Michigan).

Stone—marble, slate, and granite—constituted the other important component of Vermont mining. The first marble quarry in the state opened in 1785, with production centered around beds in Middlebury, Isle La Motte, and Swanton, as well as from Brandon to Dorset (especially in the Proctor-Rutland area). These deposits were extensive, accessible, and of very high quality. The marble was used in constructing public buildings and monuments, furniture, and gravestones or monuments. Quarrying was limited by the difficulty of transporting the heavy material, so the arrival of railroads greatly helped the marble industry, both in marketing and in connecting quarries to more central locations. By 1860, there were more than fifty quarries in the state, though the industry was centered around Rutland in the Valley of Vermont. Slate production was focused on the western slope of the Taconic Mountains in western Rutland County, spilling into New York. The first quarrying was done as early as 1812, but overall production of slate increased by 500 percent with the arrival of the railroads and a skilled workforce, primarily from Wales. Granite mining was, and still is, centered in Barre. These quarries, first opened in 1815, produce some of the finest monument-quality granite in the country, with building granite also taken from quarries in Bethel and Woodbury. Although Barre granite was used in constructing buildings such as the State House in nearby Montpelier in the 1830s, not until the arrival of the railroads did Vermont granite successfully enter the national market. Other minerals also were mined throughout the state, though they were of much less importance.

Iron working was the first significant industry in the state. Iron foundries were located throughout Vermont, though they were concentrated in the Otter Creek watershed of the northern Valley of Vermont and the Champlain Valley. After the early 1800s, though, none of these was particularly large. By the 1870s, the smaller, local foundries had started to produce specialized products for a national market. Textiles—mainly wool but some cotton—were the state's next major industry. Although this industry was important in Vermont, especially in certain towns, it was insignificant compared to the textile manufacturing done elsewhere in New England. In 1860, there were forty-six textile mills in the state employing a total of over two thousand people, nearly 20 percent of all those employed in manufacturing. The textile industry was dealt a deathblow by the competition that accompanied the railroads, which was to become Vermont's new chief industry. Other significant industries in the state included making guns, machinery, scales, and steamboats. There was much small-scale industry throughout the state as well, but most of this could not adapt to the competition that was intensified by the railroads. Overall, though, agriculture dwarfed manufacturing in the state and did so well into the twentieth century.

Since water was the chief source of power until steam came along, and since sources of water power were scattered widely throughout the state, industry also tended to be small-scale and widely dispersed. Major mills were established on Otter Creek, the Connecticut, Lamoille, Missisquoi, and Winooski Rivers, and a number of rivers in Windsor County. Floods were a real danger to these mills, which could be washed away (along with barns, bridges, and rich soil) by high waters.

The arrival of industry, combined with the growing human population, meant the beginning of pollution problems in Vermont. Industrial wastes polluted waters, as did acid drainage from copper mines. Coal burning led to localized air-pollution problems. In the largest cities like Burlington and Rutland, human and domestic-animal wastes led to public-health concerns. And the abundant mills throughout the state had numerous ecological effects—perhaps the worst of which was preventing the return of spawning fish.

In summary, the alteration of the Vermont landscape by mining and manufacturing was relatively minor compared to what occurred in other parts of the United States. Though iron was mined in much of the state, the deposits worked were small and soon abandoned. Although the stone quarries were drastic alterations, they were and are relatively concentrated and have had little legacy of chemical pollution. Only the copper mines of this period caused significant pollution problems. Vermont was substantially bypassed by the Industrial Revolution due to its relative isolation; it was distant from the population centers of southern New England and the rest of the Northeast, and transportation to the state was problematic. Roads had to cross mountainous terrain, and railroads were late in coming. Additionally, the state's low population did not provide much of a local market. So the widespread pollution problems that accompanied the Industrial Revolution in much of the Northeast were largely absent from Vermont.

Connecting with the World Beyond Vermont: Roads, Canals, and Railroads

From the colonial period through early statehood, primitive roads or paths and water were used for transportation. The first major roads were military ones: the Crown Point Military Road and the Bayley-Hazen Road, which ran from the town of Wells River on the Connecticut River to Hazens Notch in Westfield in north-central Vermont. This latter road, begun in 1776, was to have extended to Canada along the Missisquoi River to Swanton, but it was never completed. Though the fifty-four-mile road was not of military significance during the Revolution, it became an important route for settlement into northern Vermont.

Aside from these military roads, the early roads were town responsibilities. Local residents were required to build and maintain the roads, which they did not like to do. The result was few roads, poorly maintained. Vermont's topography and climate often made the roads that were built difficult to travel. Such poor roads were of limited concern when farmers were self-sufficient, but as Vermont became integrated into larger markets, better roads were needed. The initial response to these difficulties was the construction of privately built turnpikes. The first such toll road in Vermont was built in 1796, with the number of such roads peaking at over thirty statewide. Turnpikes were never popular, though, due to their tolls and poor construction, maintenance, and design (often straight over hills, making it very difficult to haul goods). People hated paying the tolls more than taxes; thus, most turnpikes were taken over by towns by the middle of the nineteenth century. The coming of the railroads spelled the final demise of private turnpikes, though a few lasted into the 1900s. In addition, a number of Vermont's bridges, especially those crossing the Connecticut River, were privately owned and charged a toll. The first—the Tucker Toll Bridge between Windsor and Cornish, New Hampshire—opened in 1796 and operated for 132 years.

As it did for the Native Americans, water played a major role in transportation in early Vermont. The existing river system and Lake Champlain were soon augmented by canals that improved and extended water transportation. One of the earliest completed canals in the United States went around Bellows Falls on the Connecticut River in 1802. Further canals on the river (at Water Quechee and Wilder) made navigation possible as far north as Barnet, though the river was often too shallow north of White River Junction. The 1830s were the high point of steam navigation on the river, with boats traveling regularly between Hartford and White River Junction. Most river transportation, though, was by flatboats driven by poles or large square sails, often moving downstream just once before being broken up for lumber. This river transportation, both steam and flatboat, was essentially killed by the railroads. On the western side of the state, the Champlain Canal connected Lake Champlain with the Hudson River in 1822, shifting trade from Canada and the north to New York and the south. The chief freight carried south was timber, followed by marble and iron. The opening of the Erie Canal in 1825 further connected Vermont to the world beyond, a world that drew away Vermonters and produced more and cheaper crops and livestock. The Champlain Canal also fueled the boom of Burlington and weakened the trade advantages of the Connecticut River Valley. Together, these Connecticut River and Champlain canals proved so successful that many more were proposed—such as ones connecting the Connecticut River with Lake Memphremagog and Otter Creek with the Batten Kill and, most ambitiously, a cross-Vermont canal. Five different routes to connect a tributary of the Connecticut to the Winooski were proposed, but the arrival of railroads killed the project for good and greatly hurt water transportation overall. A canal to the north, the Chambly, which bypassed rapids in the Richelieu River, had little effect on trade. By the time it was completed in 1843, trade had already been reoriented south, and the railroad was to arrive in Vermont by the end of the decade.

When this first railroad arrived in 1848, tracks had been laid almost to Chicago. Northern New England was something of an afterthought in the railroad boom that consumed the nation. When it came, however, it came in a flurry, with nearly five hundred miles of track laid in Vermont by 1855 (see map 4.2). The three main systems were the Vermont Central, which ran from White River Junction through Northfield to Burlington; the Rutland and Burlington, from Bellows Falls to Rutland and on to Burlington; and the Grand Trunk, which connected Montreal and Portland through the Northeastern Highlands. Further lines were soon laid connecting other major cities and towns in the state to these main routes. This basic network, augmented by numerous spurs for logging and moving granite and marble, was completed by 1870.

Vermont did not have the population or industry to make railroads profitable; the key was through trade from Canada and the Great Lakes to Boston and Portland on the Atlantic. This did not lessen the tremendous effect these railroads had on Vermont, however. Most important, they fully

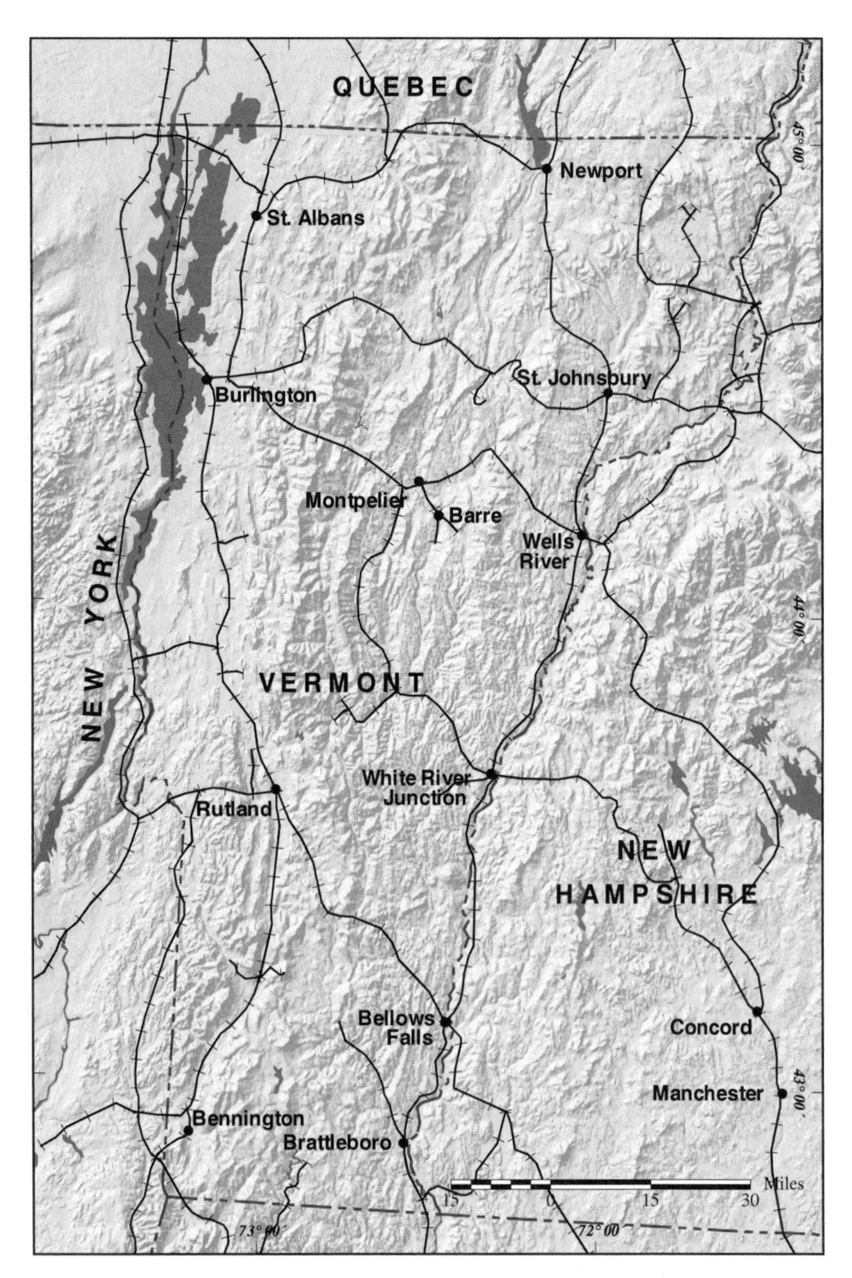

MAP 4.2. Railroads in Greater Vermont, 1850–1900.

integrated the state into the regional, national, and international economy and society. They cut marketing costs almost in half, making it easier to exploit such raw materials as copper, granite, marble, and slate. As mentioned above, the railroads also stimulated the lumber industry and the dairy industry. The speed of trains, as well as the lower transportation costs, helped Vermont become a major dairy supplier—first for butter and cheese and later for fluid milk—for the urban markets to the south. Railroads also helped some manufacturing (such as machine tools, scales, and wood products), but hurt or killed much small-scale manufacturing (such as ironworks, tanneries, and textile mills) that could not survive the new competition brought by the rails.

A second, equally significant effect the railroads had was to alter certain patterns within Vermont. Towns along rail lines grew in population and wealth; those not on the lines declined. Hill towns, in particular, withered further during the railroad era. In addition, towns shifted their centers downhill or from one location to another in response to the routes of the tracks. Among the towns that benefited most from the railroads were Northfield, Rutland, White River Junction, and Windsor. Burlington, on the other hand, suffered somewhat by losing much of its backcountry market, by being bypassed as most rail traffic went through Essex Junction, and by the decline of Lake Champlain traffic. Other lake ports were hurt even more, and their docks and warehouses fell into virtually complete abandonment.

Land values along or near the lines increased dramatically; indeed, the total value of Vermont farms increased by one third in the 1850s. The railroads increased mobility and communication (telegraph lines accompanied the rail lines). They made it easier for emigrants to leave Vermont for cities or the West, yet they also helped spawn tourism as a major commercial activity. Wooden-plank roads enjoyed a brief boom around this time, connecting towns without rail to those along the tracks. The interest in such roads, like the one connecting Bristol and Vergennes, lasted about as long as the planks, roughly ten years.

During a period dominated by subsistence farming, good transportation systems do not matter that much. But as commercial agriculture began to command more attention, and mining and manufacturing arose, better transportation was an economic imperative. The Champlain Canal and the railroads served to incorporate the human economy of Vermont into the world beyond its borders, and Vermonters gradually lost control over what happened to them and to the landscape of their state.

Roughly one hundred years after European settlers began to stream into Vermont, the state was a landscape transformed. The native peoples were largely gone, old-growth forest had virtually disappeared, large animals such as the elk, the mountain lion, the wolf, and the wolverine were gone or virtually so. Instead, there were over 330,000 people, nearly 600,000 sheep —down from a peak of over 1.6 million—and 180,000 dairy cows. But these changes were not permanent; rather, the changes of this period established the extremes—thus far—in how humans have altered the landscape. In the previous chapter we focused on the importance of establishing the borders of Vermont. By 1870, those borders had become less and less meaningful in the face of an increasingly competitive market economy, served by an expanding rail system. Vermont had no control over the price of wool or the availability of land in western states. Those factors, more than decisions made on individual farms in the state, set the stage for the next era of change on the Vermont landscape.

5

The Return of the Forest

The Vermont Landscape from the Civil War through 1950

FROM THE 1870s through the middle part of the twentieth century, Vermont was generally viewed as being in decline. But like most views, this one depends on perspective. The boom period of the previous century, when over 300,000 Europeans or European Americans moved into Vermont, transforming the state from wild forests to farms and woodlots, came to a slow end. It was followed by nearly a century of greatly reduced population growth, of continual agricultural specialization, and of largely missing out on the Industrial Revolution, which was transforming the landscape and society of so much of the rest of the nation, especially nearby southern New England. Viewed from another perspective, though, it was a time of recovery. Much of the land cleared by settlers slowly returned to forest. The conservation movement began, leading to the establishment of programs to manage uses of the state's forests, waters, and wildlife, and to the purchase of lands by governments at all levels for a variety of conservation purposes. The Vermont landscape remained dynamic, but during this period the general trend of change was itself altered: Humans went from clearing the forest to actively and passively aiding its return.

Farming Declines, Forests Begin to Recover

The three main trends in Vermont agriculture from 1870 through 1950 were the continued decline of general farming in the state; a correlated reduction in the amount of cleared land in the state; and the further specialization of remaining farms into dairy production, especially fluid milk for the Boston and New York City markets. This decline was most pronounced from the end of the Civil War period through the turn of the century. During that time, Vermont's wheat production fell 92 percent and corn production fell 30 percent. Sheep farming remained in decline. This was not a function of low productivity; indeed, as late as 1889 Vermont was the leading state in terms of yield of corn per acre and third in yield of oats per acre. These production declines continued into the twentieth century: from 1900 to 1930, corn production declined nearly 90 percent, buckwheat 88 percent, sheep 83 percent, and oats over 60 percent.

The main reasons for these declines were essentially the same as they had been earlier: the lure of the farmland to the west, the draw of the cities, and the decreasing attractiveness of the hill farms. To the west, a new comparative advantage beyond cheaper land and more productive soil became more and more important. As agriculture became increasingly mechanized, Vermont farmers could often only dream about making use of the big new machines, which were too large for their too hilly and too small farms. To the west, the land was flat, and larger farms were easier to establish. The resulting productivity edge of western farmers, combined with cheaper transportation of western products via railroads, squeezed Vermont farmers economically.

Cities of southern New England also grew increasingly attractive to Vermonters in the latter third of the nineteenth century. The best western lands had been settled, and cities were becoming the economic centers of the nation. As the most ambitious and most talented young people left the farms for the cities, rural life became more isolated. Many older farmers had hard times keeping their farms running without the help of their children, leading to many abandonments.

Finally, those farming in the hills found their lands to be the least competitive in a commercial agricultural setting: The soils and climate compared unfavorably to the valley farms; they were farther from rail connections; and new machinery often could not be used on the hills. None of this had mattered when Vermont farming had been focused on self-sufficiency. By the late 1800s, though, these characteristics mattered a great deal. Thus hill farms typically were the first to be abandoned. Complementing these major factors underlying the decline of farming in Vermont were its harsh climate, declining soil fertility, and an often unfair tax burden on farmers. There were several more specific causes of the continuing wool decline (see table 5.1). Two of these—competition from Australian wool and a decline in the wool tariff in the 1880s—underscored the inability of Vermont farmers to control their own destinies. Another reason illuminates the futility of efforts to control nature: Decades after killing off the wolves to protect livestock, sheep farmers faced a constant loss in their flocks due to attacks by wild dogs.

Not surprisingly, this decline in farming led to a decline in cropland and pasture in Vermont (see table 5.2). The percent of land in the state classified

TABLE 5.1.
Human and Animal Populations in Vermont, 1880–1950

Year	Human population	Density (sq. mile)	Sheep population	Cow population
1880	332,286	35.8	439,870	217,033
1890	332,422	35.8	333,947	231,419
1900	343,641	37.1	296,576	270,194
1910	355,956	38.4	118,551	265,483
1920	352,428	38.0	62,756	290,122
1930	359,611	38.8	51,175	264,000
1940	359,231	38.7	35,946	279,141
1950	377,747	40.7	12,000	261,370

TABLE 5.2.
Farmland and Forestland in Vermont, 1880–1950

Year	Farmland (acres, % of state)	Forestland (estimated acres, % of state)
1880	4,882,588 (82)	2,270,900 (38)–2,383,400 (40)
1890	4,395,646 (74)	n/a
1900	4,724,440 (80)	2,552,500 (43)
1910	4,663,577 (79)	2,500,000 (42) [1909]
1920	4,235,811 (71)	n/a
1930	3,896,097 (66)	3,562,000 (60)
1940	3,666,835 (62)	3,549,000 (60) [1938]
1950	3,527,381 (59)	3,729,700 (63) [1948]
		3,845,900 (65) [1952]

Note: The estimates of forestland in Vermont (and elsewhere) are extremely rough until around 1950. Hence, these figures are at best approximations. Also, there is an overlap between farmland and forestland due to a significant amount of farm acreage in forest (usually referred to as woodland).

as improved farmland peaked at 68 percent in 1870. By 1900, this figure had declined to 45 percent. The decline continued into the 1900s, with the acreage dropping from 2.1 million acres in 1900 to 1.4 million acres in 1930. The number of farms in the state followed a similar pattern, peaking at over 35,000 in 1880, falling to 33,000 in 1900, and to fewer than 24,000 in 1940. Some of these farms were combined with neighboring farms, leading to an increase in average farm size to 155 acres in 1940 (the largest in New England). Nevertheless, much of the land went out of cultivation; by 1940, the percent of farmland reported as unimproved had risen to 60 percent. This decline in improved farmland was most concentrated in the hilly regions of the longest settled places. Farm abandonment was most heavily concentrated in Windham, Windsor, and Bennington Counties. These problems affected all farmers, though, in that they reduced the value of farm property.

The one segment of Vermont agriculture that did well during the late nineteenth and early twentieth centuries was dairy. Farmers took advantage of Vermont's proximity to Boston and New York City to supply them with perishable dairy products that their western competitors were unable to deliver. As dairy farming became more commercial, cheese factories and creameries began to replace home production of cheese and butter. Butter production in creameries rose from 5,000 pounds in 1879 to 22 million pounds in 1899 (with an additional 18 million pounds made at home). Vermont led the nation in butter production at this time, and Saint Albans was home to the world's largest creamery. Cheese production peaked in the mid-nineteenth century, steadily declining thereafter—since cheese could last almost indefinitely, Vermont farmers had little market advantage for this product. The market for fluid milk developed more slowly since it remained difficult to transport to the southern New England cities before it spoiled.

These changes in Vermont farming are borne out by reports by state officials and income statistics. The state agricultural report for 1891–1892 notes that "we are now in a transition state, passing from the system of extensive farming to that of intensive farming." Prior to 1870, the typical Vermont farm family made its living through general farming; by 1900, half of Vermont farms received most of their income through dairy. This transformation was not as easy as it sounds since many farms were ill suited to dairy; those farms that could not adjust were those most quickly abandoned.

The surge in dairy production ended around 1900, with production declining over the next decade. The cause was once again competition from the west. Changes in technology (mainly in terms of better refrigeration) allowed midwestern producers to supply butter and cheese to northeastern cities for less than could New England farmers. In response, butter and cheese production in Vermont fell by over half from 1900 to 1920. These technological changes, unlike many previous ones, did provide a benefit for Vermont farmers: They could now start to sell fluid milk in cities further to the south. By 1930, Vermont dairy production had recovered to 1900 levels thanks to the boom in the fluid-milk market. Boston began to receive milk from southeastern Vermont just before the turn of the century, and by 1920 farmers throughout the state were part of the Boston milkshed. By the end of the decade, 50 percent of Boston's milk came from Vermont. Some western Vermont farmers became part of the New York City milkshed. Overall, 85 percent of the milk produced in Vermont in 1930 was shipped out of the state. This fluid-milk market remains the foundation of Vermont dairy farming today. In the 1920s, the state led the country in dairy production per capita and had a greater reliance on the dairy industry than any other state. The state's commitment to dairy mirrored the sheep-boom years of the middle nineteenth century, and, as was the case during the sheep boom, the farmers of the state were fully integrated into a market over which they had very little control.

Vermont's competitive advantage in milk production did not last. Midwestern farmers could produce milk for less than Vermont farmers because of a longer pasture season and their ability to grow or acquire less-expensive feed. These differences had already cost Vermonters the butter and cheese markets. Changes in Vermont's dairy industry were evident in the first half of the twentieth century, as the percent of farms with cows fell (from 87 percent in 1920 to 67 percent in 1945), and the cows per farm rose (from over eleven to over fifteen). As productivity became more important, the retreat from marginal farmland accelerated since better land was needed for pasture and to grow feed for the herd. By the 1930s, the dairy industry had begun to focus on those counties with more productive land, such as Franklin, Grand Isle, and Chittenden. Increased capital investments, in the form of specialized breeds and silos to store winter feed, also became much more common in the early twentieth century.

Other smaller specialty crops also began to flourish in the later 1800s and early 1900s. Vermont farmers had a market advantage for eggs similar to that for dairy products, and Vermont egg production rose from 36 million in 1879 to 80 million in 1929 to 180 million in 1950. Overall, though, the Vermont farm economy was less reliant on poultry than was that of any other New England state. A commercial apple market in the cities led many farmers to rescue neglected orchards or plant new ones. Maple sugaring also was rejuvenated. The Vermont Maple Sugar Makers Association, formed in the 1890s, sought to improve quality, increase production, and better market Vermont maple-sugar products. It was successful, with production more than doubling to over 12 million pounds by 1930. There was also a shift in final product: In 1840 nearly all of the sap was made into maple sugar; by 1940, it was almost all made into maple syrup. Over 3.6 million acres were tapped in 1940, the highest figure of any state. Indeed, Vermont had nearly 40 percent of all the trees tapped in the United States. In some places, fern picking in the woods—for city florists to use in their shops and at Christmas—brought in extra money.

As Vermont farmers became part of increasingly integrated markets, the less control over their own destinies they had. The creation of margarine in 1870 by a Parisian chemist or the rate schedule the railroads charged to ship milk to Boston, for instance, could have as much effect on Vermont farmers as a drought or a cold winter.

To lessen the effects of a completely free market on Vermont farming, and to try to stem the flow of farmers from the state, the state government initiated an activist agriculture policy. The Vermont Board of Agriculture was founded in 1870 to help spread improved farming techniques. The board was helped by fairs, a new agriculture college at the University of Vermont (1865), and an agricultural experiment station (1886). Although the information the board provided helped many farmers, ironically it also hastened the exodus from the farms as productivity levels on the remaining farms went up. Farmers began to seek governmental aid in their efforts to compete in the marketplace. State laws, for instance, sought to protect sheep from diseases; national laws, such as tariffs, sought to protect wool, dairy, and maple-sugar producers. The Vermont Board of Agriculture made efforts to attract immigrants from other states, and especially other countries, to settle the farms that had been abandoned. The chief focus was on attracting Swedes, but only twenty-seven families came from Sweden. Most of the new farmers were unsolicited French-Canadians. Private groups formed as well: the Vermont Dairyman's Association in 1869, the Vermont State Grange in 1870, and the Vermont Farm Bureau in 1918.

The federal government began to get involved more actively in actual farming with the establishment of county extension agents in 1914, but it was the Depression that really brought Washington fully into agriculture. A host of federal agricultural programs—such as the Agricultural Adjustment Administration, the Farm Security Administration, and the Soil Conservation Service—came to Vermont to help financially pressed farmers. Federal efforts to regulate and stabilize dairy supplies and prices began during the Depression era as well.

Agriculture in middle twentieth-century Vermont was drastically different from agriculture in middle nineteenth-century Vermont. By 1930, nearly 75 percent of Vermont farms were categorized as specialized (meaning that one product accounted for more than 40 percent of farm income), mainly in dairy. Six percent of the farms in the state were self-sufficient. One hundred years before, nearly all farms were general ones that were virtually self-sufficient. As Vermont agriculture became increasingly connected to larger market forces, small self-sufficient farms disappeared, and the remaining farmers relied on freshness to provide a competitive advantage in nearby urban markets. These changes in agriculture had two landscape effects, one direct and the other indirect. The direct effect was the return of forests as land went out of cultivation. The indirect effect was the stagnation of human population growth and the alteration of its distribution in the state.

From 1880, the chief change to the Vermont landscape was that it began to revert to forest—the complexity and diversity of which will be more fully discussed in chapter 7. There is no documented figure for just how low the percentage of forest covering the state's landscape dropped. The Vermont Department of Forests, Parks and Recreation estimates that the state was only 25 percent forested around 1850. Other published estimates put the low of forest covering at roughly 38 percent of the state in 1880. Regardless of which figure is more accurate, it is clear that the trees returned rapidly in the latter part of the nineteenth and early twentieth centuries, reaching 60 percent coverage by 1930 (see table 5.2).

This returning forest, though, was not left undisturbed. Far from it; Vermonters continued to take what resources were valuable from the woods. Into the 1920s, for instance, over half of farm woodlands were grazed (a figure that has gradually declined since). A major charcoal industry, feeding large kilns, flourished from 1860 to 1900. Harvesting for lumber reached a secondary peak of 384 million board feet in 1890 (75 percent of which was spruce). Burlington's role as a major sawmilling center declined dramatically in the latter part of the nineteenth century, due in part to the duty on imported Canadian lumber in 1897 (imports fell by two thirds during the first year). Soon after, New England and Vermont generally ceased to be important factors in the national lumber-supply picture. By 1925, the Vermont cut had fallen to 91 million board feet, since most of the best trees had been taken. Through the 1930s, harvests averaged 125 million board feet, less than half the average from 1900–1909. An upturn began in the 1940s with the rise of active forest management. Annual harvests averaged over 200 million board feet, reaching 342 million board feet in 1946.

There were also a number of changes in terms of species harvested and goods produced. With the mature stands of white pine and spruce gone, not to return for many years, the focus of the lumber industry shifted from softwoods to hardwoods. By the late 1920s, 61 percent of the harvest was hardwoods, mainly maple and birch. Veneer mills to make use of these hardwoods started to appear in Vermont by 1880. Furthermore, a new process to make paper from wood pulp was developed after the Civil War, creating a further source of demand for spruce and fir. In the late teens, there was a surge in demand for wood for heating, as coal was diverted to the war effort. Finally, a market for Christmas trees began in the early 1900s.

Despite these new and continuing demands for forest products, a number of factors combined to lessen the pressure on managing the returning forests for commodities. First and foremost, in addition to the decline of improved farmland in Vermont, was the decline in demand for wood for fuel. Coal, oil, and electricity all helped to keep more trees in Vermont's forests. For instance, the Vermont Central Railroad decreased its use of fuelwood from 170,000 cords in 1874 to less than 70,000 cords thirteen years later. A second factor was that per-capita consumption of lumber in the country peaked in 1906 as other building materials began to substitute for lumber. These factors, combined with increasing national and international trade in forest products, meant that Vermont's—and New England's—forests did not have to supply the needs of the people who lived there. Vermont's forested landscape began to recover not because fewer people lived there but because the needs of the people in Vermont for food, fuel, and shelter were increasingly being supplied from somewhere else. Cutting fewer trees to heat homes in Vermont meant mining more coal in Pennsylvania to serve that purpose.

The return of the forests affected the wildlife on the landscape as well. Deer, virtually extirpated from Vermont, were reintroduced by private sportsmen in 1878 and began to steadily increase in numbers. Edge and open-field species, such as skunks and raccoons, also thrived during the transition from farmland to forest (and still thrive in today's fragmented regenerating woods). Problems came, too, notably in the form of diseases such as the chestnut blight and the white pine blister rust. Each of these forest subjects—wildlife, fragmentation, and exotic diseases—as well as others, will be more fully examined in chapter 7.

As anyone who has spent much time walking in Vermont forests knows, the return of these forests has not completely erased the traces of human habitation that directly preceded them. The vegetation grows back quickly and from afar covers the traces of human activity, yet on the ground old mills, ironworks, stone walls, apple orchards, roads, and foundations can still be found. In these woods, the distinction between nature and culture is blurred.

The Rise of Conservation and Public Lands

Conservation, centered on the prudent management of natural resources—especially wildlife, forests, and water—became a significant national movement in the late 1800s. Vermonters were very much a part of this movement. By the 1920s, the federal government was significantly involved in the management of the resources of Vermont and all other states, a process that was often quite controversial.

A number of Vermont towns had passed wildlife laws prior to independence. In 1762, Bennington elected a deer rift (also called a deer reeve) to enforce laws against killing deer out of season. A number of other towns followed suit, but by the early nineteenth century deer were largely gone from southern Vermont, and so were the deer rifts. Newbury established a wolf bounty in 1785. Vermont's first legislature, in 1779, passed wildlife laws as well: There would be no deer hunting from January 10 through June 10, and bounties were established on wolves and catamounts, more widely known as mountain lions. The last wolf bounty was paid in 1902, and it was removed from the books in 1904; the last catamount bounty was paid in

1896. During the 1800s and early 1900s, bounties were also established on bears, foxes, porcupines, rattlesnakes, and wildcats (bobcats and lynx).

By the 1820s, fish stocking was taking place at the local level, with some efforts at regulation by local governments. The towns were ill equipped to enforce these rules, however, and the state legislature soon became involved. It passed many laws regulating fishing in particular places, but they did not work. A more systematic approach came in the 1850s. In 1857, George Perkins Marsh was appointed fish commissioner, and he wrote a report describing the decline of fish species and populations within the state. The first institutionalized state response to poor natural-resources management came in 1866 when the state Fish Commission was established. The three fish commissioners cited numerous problems affecting the state's fisheries: deforestation causing increased water temperatures and sedimentation, as well as dried-up streams; dams, mill wheels, and pollution due to industry; and too much fishing, much of which was done at the wrong time of year. Chapter 9 contains a more detailed discussion of the ecology of these aquatic communities and how they have been affected by humans.

A statewide fishing code was passed in 1867, which included closed seasons for some species and regulations on certain types of fishing equipment. This law, like the local regulations and site-specific state laws preceding it, was widely ignored. People had been used to seeing fish and game as free common resources and were very skeptical of state regulatory efforts. To address some of these problems, a system of integrated town, county, and state wardens was established in 1876, laying the groundwork for all future wildlife management. The fish commissioners also worked to secure interstate cooperation in an unsuccessful effort to restore the salmon and shad runs, which had disappeared from the upper Connecticut River watershed because of dams in Massachusetts and Connecticut.

Meanwhile, the deer population continued to decline, and deer hunting in Vermont was against the law from 1865 to 1897. During this time, existing deer herds recovered, and some successful efforts to reintroduce deer to other parts of the state occurred. The legislature broadened the jurisdiction of the Fish Commission in 1892, renaming it the Fish and Game Commission. Not long after, deer hunting was allowed again; hunting licenses were required—first for nonresidents and then for residents as well.

During the last two decades of the nineteenth century and the first decade of the twentieth, the foundations of Vermont forest policy were laid down. The legislature appointed a committee in 1882 to study the forest situation in the state; ten years later, a forestry commission was established. In 1894, Governor Urban Woodbury included management of the forests as a topic in one of his messages to the legislature: "Owners of timber lands in our state are pursuing a ruinous policy in the method used in harvesting timber. . . . The value of our water powers and the attractiveness of our scenery and the preservation of game and fish also call for reform." In 1904, the state Board of Agriculture was directed to appoint one of its members as Forestry Commissioner. (The private-sector Forestry Association of Vermont was founded the same year.) Four years later, the board was authorized to hire a professional State Forester to enforce forestry laws and manage the state-forest system, which was established the following year. During this period of the birth and growth of the forestry profession in the United States, its two major national figures—Bernhard Fernow and Gifford Pinchot—both spoke in Vermont.

The state forester and Forest Service sought to encourage good forest management on private lands. In 1911, the state forester began work with private landowners who were interested in forest management. Much of the stimulus for these state actions came from two federal laws—the Weeks Act (1911) and the Clarke-McNary Act (1924)—that established programs and provided resources for states to work with private forestland owners. Later, in 1945, a law to further this aim was passed, entitled "An Act to Assist Forest Owners and Operators in the Promotion of Maximum Sustained Productivity of the Forests, to Disseminate Information Relative to Forest Practices." This early period culminated in 1915 when the legislature passed a law that allowed towns to acquire and manage forests and allowed for the creation of school forests, to be managed by the state forester, to support local schools. This law set the stage for municipal forests in Vermont (discussed below). Although the Forestry Association disbanded in 1922, other groups have arisen since to take its place. A state Forest Service, supervised by a Commissioner of Forestry, was created in 1923 to further the work of reforestation and forest conservation. This mission was expanded to include controlling forest diseases and fires. The latter had been a significant concern because of major fires in the early twentieth century, but thereafter it became a minor problem in Vermont. For example, from 1936–1941, the state had 229 fires that burned an average of thirteen acres each. Forestry in the state got a real boost from the federal Civilian Conservation Corps during the 1930s. Eighteen camps with over 3,500 men worked on a variety of projects, including tree planting, road building, pest and disease control, trail cutting, and shelter construction.

A final component of the forestry program was reforestation. Through the early twentieth century, the return of the forest to Vermont was based almost wholly on natural regeneration (afforestation). A shift to reforestation began in 1906, when a state nursery was established. By 1948, over thirty million trees from it had been planted on state and private lands throughout Vermont.

State capacity expanded in other areas as well. In 1886, a Board of Public

Health, which investigated sanitation and epidemics, was created. In 1935, a Department of Conservation and Development—which included the newly created Fish and Game Service and the Forest Service, among other agencies—was formed. This department's name was changed to Natural Resources in 1943, but it lasted only four years, when it was replaced by a Department of Recreation, an again-independent Fish and Game Service and Forest Service, and a new Water Resources Board.

The early twentieth century also marked the beginning of significant state and national conservation of public lands. The tradition of public lands in Vermont and New England can be traced back far into the colonial past. Common land, for the use of all the town, was first established before the mid-1600s. These lands, designated by the town proprietors, could be used for grazing and firewood collection. Such lands were never significant in Vermont, though, due to the abundance of forestland and the different role for proprietors (see chapter 3). Another type of colonial public land has played a significant role in Vermont: specific parcels set aside to support churches and schools. These lands, originally totaling as many as 300,000 acres, were leased to generate income for these institutions or used as building sites for churches or schools. In many cases, these lands did not supply enough income, so the state gave a school an additional parcel to supply income or fuelwood. Until 1937, such public lands created by Wentworth (New Hampshire) or Vermont charters could not be sold; the Vermont courts ruled that these lands were to be held in trust for future generations. The revenue from these lease lands was collected by either town select boards, the Episcopal diocese, or the University of Vermont. Since the 1937 law allowing the sale of such lease lands (with the proceeds to go into a relevant trust fund), many of these lots have been sold, but many have not. In the latter case, select boards liked the return (even though it is quite low) and leasees liked the low rates.

Many towns took advantage of the 1915 town forest law in order to generate revenue and jobs for the town, to reclaim land, to stimulate wood-products firms in the area, to protect water supplies, to provide recreation, or to serve as a memorial to war veterans or deceased family members. These town forests were acquired in a variety of ways. Some lease lands that had become delinquent were converted to town forests. In some locales, the town poor farm became a town forest. There was also the occasional gift of land, but most of these forests were established by purchase. In one interesting case, Sheffield bought two remote hill farms and reforested them to save on the high costs of keeping the road to these farms open in the winter.

By 1930, there were forty-two town forests encompassing almost nine thousand acres. A surge of new town forests occurred after 1945 when the original law was amended to require the state to reimburse the town for half the purchase price of town forestland. Further beneficial legislation was passed in 1951, requiring towns without a forest to include articles proposing them at town meeting. The state Department of Forests and Parks sought to establish a forest in every town. Following these changes, the number of town forests increased quickly. In 1950, sixty-eight towns had forests totaling over 16,000 acres. By this time, the uses of these forests had shifted as well, moving away from commercial production toward public recreation. Another shift was the increased use of municipal watershed plantations to protect town water supplies. These plantations were primarily softwoods, which did not generate as much organic matter in the reservoirs and provided quick-growing, high-quality timber. In many instances, town watershed lands were located in other towns; Rutland, for example, owned over four thousand acres, mainly in Mendon. Indeed, two Massachusetts towns still own over four thousand acres of watershed forests in southern Vermont.

The final component of town forests was the preservation-oriented forest parks. As preservation was arising on the national scene in the late 1800s and early 1900s, so too was it becoming important in New England and Vermont. Here, the grand pristine landscapes like Yellowstone and Yosemite did not exist, so the focus was on smaller forest parcels for recreation, spiritual contemplation, and nature study. Never part of an organized campaign, these forest parks were typically donated to a town, as Hubbard Park was in Montpelier or Battell Woods in Middlebury.

The state forest system was established in 1909 with the aims of stimulating private forestry by example, of protecting water sources, and of raising quality timber. The state purchased 450 acres for the first forest in Plainfield. During the 1910s, twelve more state forests totaling nearly 30,000 acres were established, including one by donation that encompassed the summit of Camel's Hump. By 1950, twenty-four such state forests existed, though the state Forest Service envisioned a system of 300,000 acres and a town forest in every town that had suitable land. The state park system (technically referred to as forest parks) was chartered legislatively in 1929, five years after the first such park was donated at Mount Philo. Twenty forest parks were established by 1950.

Efforts to establish a national forest in Vermont began in 1905, when Marshall Hapgood offered to sell the federal government a large tract of land in the Green Mountains. A few years later, Joseph Battell explored donating some of his land for a similar purpose. In 1911, the Weeks Act was passed, which permitted the federal government to buy forestland to protect navigable waters. This law allowed for the creation of national forests in the east. Through the 1910s, plans for a national forest in Vermont were put on hold as the national Forest Service worked on establishing the White Mountain National Forest in New Hampshire and the Allegheny National Forest in Pennsylvania. A 1920 Forest Service study identified two areas in Vermont as meeting the Weeks Act requirements for potential national-forest designation: the Nulhegan or northern unit in the Northeastern Highlands, totaling 240,000 acres, and the southern unit in Windham and Bennington Counties, totaling 100,000 acres. Despite the study, a Vermont national forest was not high on the priority list, given forest-conservation needs—mainly to prevent erosion and protect watersheds—elsewhere in the country.

The 1924 Clarke-McNary Act extended the kind of land the Forest Service could acquire for national-forest lands to essentially any land. This changed the situation in Vermont, since the proposed national forest would not have to be tied to navigable waters. The following year, Vermont passed the Enabling Act of 1925, granting the Forest Service the approval of the state to purchase land anywhere in Vermont for a national forest. Supporters of tourism and the forest-products industry both felt such a forest would serve the state well.

The catalyst to move ahead on a national forest was the 1927 flood (discussed below). In order to help prevent such floods in the future, more lands needed to be kept as forest, and the state forest system could not handle this alone. In 1928, Vermont made an official proposal to the National Forest Reservation Commission for a Vermont national forest consisting of a northern and southern unit of roughly 300,000 acres. The Commission reviewed the proposal, then made its own recommendation: a 370,000-acre national forest entirely in the southern part of the state, with federal land purchases totaling no more than 100,000 acres. This southern national forest would focus on watershed protection, timber, and recreation, needs that were not as pressing in the proposed northern unit.

Over the next few years, the size of the national forest was adjusted downward, first to 250,000 acres in the four southernmost counties in 1929, then to 100,000 acres in those counties, eliminating the southernmost part of the proposal. In 1932, the Green Mountain National Forest became official with a proclamation by President Herbert Hoover and the purchase of the first land. The official boundaries included 102,000 acres, with plans for 90 percent federal ownership within this area. Throughout the 1930s, the boundaries of the national forest were further adjusted. In response to pleas by conservation leaders that a 100,000-acre national forest was not large enough, a new northern unit in the central part of the state was approved in 1935. This 205,000-acre unit would eventually include 175,000 acres under federal ownership. That same year, the deleted southern portions of the national forest were reinstated. Overall, the Green Mountain National Forest proclamation boundaries then stood at 580,000 acres in forty-five towns, covering 9 percent of the state.

Not all Vermonters favored these expansions. Indeed, as the federal government began to buy lands, some expressed alarm over lost property taxes and growing federal control. This latter fear was only exacerbated by the Green Mountain Parkway controversy (discussed below). In 1935, Vermont amended the Enabling Act, requiring that Forest Service land purchases be approved by a new State Land Use Board. Further amendments two years later required written approval by the relevant town select board before the State Land Use Board would approve any national-forest purchase.

Despite these changes, acquisition and management of the new national forest moved ahead. Acquisition was most successful in the first five years of the forest's existence, when more than 160,000 acres of land were acquired. Rising property values and World War II led to decreased acquisition; fewer than eight thousand acres were acquired between 1939 and 1949. Management on the lands focused on developing recreation sites and improving the quality of timber on the lands, chiefly quality hardwoods in existing forests and softwoods on abandoned farmland. Increased timber demands during World War II led to an increase in the cut on the national forest from less than 1 million board feet in 1939 to 8.5 million board feet in 1945. In the later 1940s, the Green Mountain National Forest was the highest revenue-producing forest for its size in the east. This trend of relatively high timber harvests on the forest continued into the late 1960s.

In 1940, public lands in Vermont totaled more than 240,000 acres: more than 160,000 federally owned, more than 70,000 state owned, and 10,000 municipally owned acres. About this time, the New England Planning Commission was recommending that roughly 15 percent of the New England landscape should be publicly owned—3 million acres by the federal government and 3.5 million acres by the states. At just barely 4 percent, Vermont still had a long way to go to meet this goal, but federal, state, and local governments have continued to acquire more public land (see chapter 6).

Although the establishment of the Green Mountain National Forest was a success, it was soon followed by three examples of activist federal conservation gone too far for Vermonters. The Green Mountain Parkway, proposed in 1933, was to be a 250-mile road through a thousand-foot-wide corridor of protected land, connecting state parks and terminating in the 125,000-acre Green Mountain National Park, running from Mount Ellen in the central Green Mountains to the Lamoille River Valley to the north. The parkway was to travel the summits in some places, dip into the valleys in others. Some advocates suggested that the parkway corridor might be expanded to a width of five to fifteen miles, creating a total project of one million acres. The parkway, modeled on the Blue Ridge Parkway in Virginia and North Carolina, was to be the northernmost piece in a system beginning in Georgia. The goals of this system were to reduce unemployment and lay the foundation for future economic development based on tourism.

Vermonters were immediately split over the proposal. Supporters thought that the Green Mountains were going to be developed sooner or later, and such development would be handled better by the National Park Service than by commercial interests. Opposition was led by the Green Mountain Club, especially its out-of-state members, which wanted to keep the mountains wild. The club was joined in opposition by national conservation leaders, including Aldo Leopold. In a 1936 letter to the Rutland *Herald*, he wrote: "There seems to be something approaching an epidemic of expensive unneeded roads invading the last remnants of wild country still available in the United States. . . . I can assure you that any desire on my part to revisit the Green Mountains would be forever canceled and destroyed if your state goes ahead with this road."

Other opponents did not like the idea of federal control of more land in the state, nor Vermont's share of the cost ($500,000 of an $18.5 million project). The Vermont House voted down an act allowing the federal government to acquire a 50,000-acre corridor 126–111 in 1935 (the Senate had narrowly passed the bill), and the following March the parkway lost again in a town-meeting referendum, 43,176 to 31,101. The project was dead. A proposal to create just the Green Mountain National Park the following year went nowhere.

The second example of federal, centralized conservation gone too far for Vermont involved the proposals of the Resettlement Administration. This agency sought to move people off marginal lands throughout the country, where, it argued, these people could never make a proper living. Instead, they would be resettled in cities or on better farmland. Bureaucrats from the agency arrived in 1934 with plans to move people from 19,000 acres of marginal lands in four areas in central and southern Vermont. The federal government would purchase the land, and, after it had been reforested, it would be transferred to the state, with the stipulation that it could never be inhabited in the future. A general plan suggested 55 percent of Vermont might eventually come under federal ownership through this program because of the marginal productivity of Vermont lands. The opposition to this program was quick, loud, and widespread, led by then-Governor George Aiken. (Aiken, a wildflower enthusiast, was aided greatly in his campaign for governor by giving illustrated wildflower talks throughout the state.) Due to this opposition, no land in Vermont was ever acquired by the Resettlement Administration.

The final issue involved perhaps the most significant natural event in Vermont's history: the flood of November 1927. This was more destructive

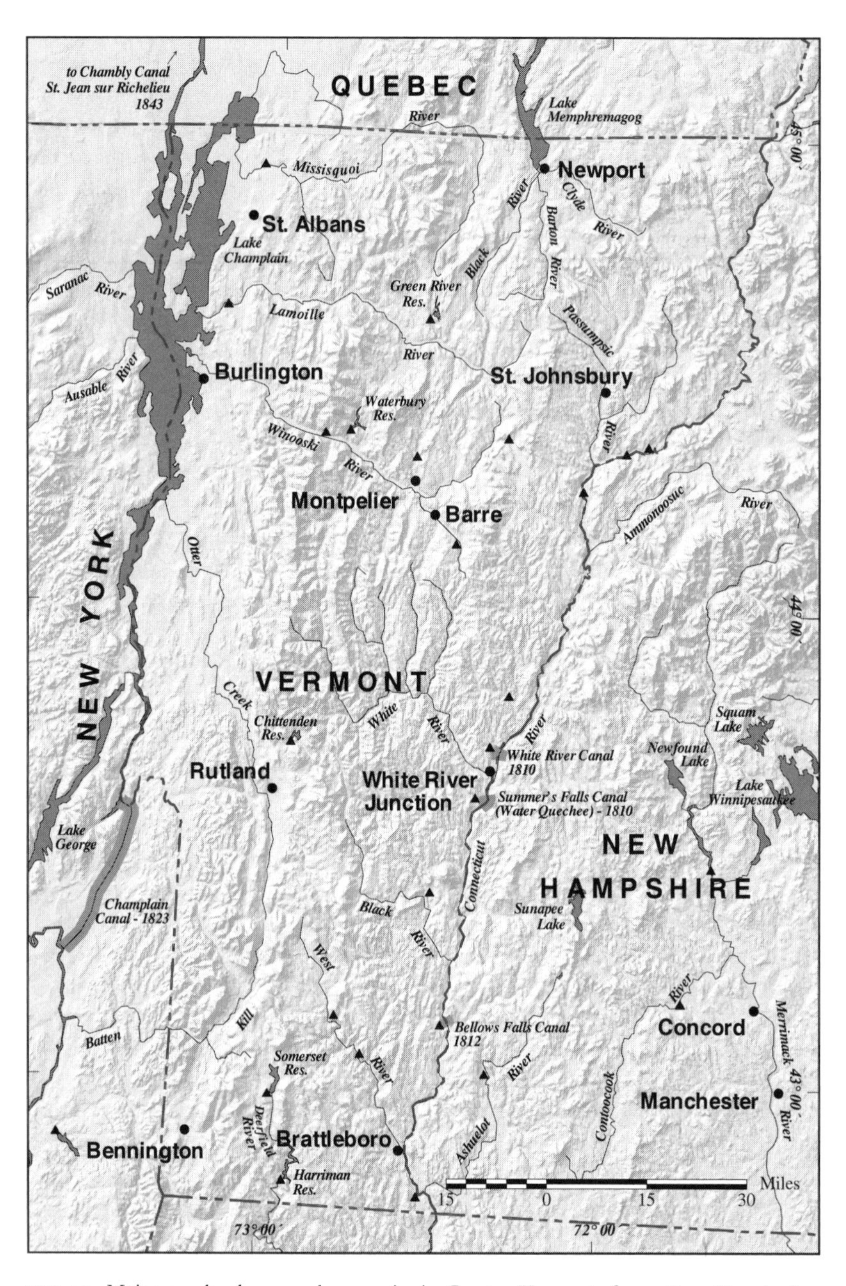

MAP 5.1. Major canals, dams, and reservoirs in Greater Vermont, from 1802. *Key:* ▲ dam; ▬ canal.

than any Vermont flood in post-Columbian history, with the floodwaters in the Connecticut River Valley at their greatest height in three hundred years. Although all of New England was affected by the floods, Vermont was hardest hit: eighty-five people died, and there was massive damage to roads and railroads throughout the state. According to the commissioner of agriculture, "Nothing had occurred during the history of the state which had dealt such a staggering blow to the agricultural industries of Vermont."

After the flood, Vermont and the rest of New England started to focus seriously on flood-control projects. Vermont planned a network of private storage reservoirs to deal with the problem. The federal government, though, had a larger vision. In 1928 it unveiled a master plan of eighty-five dams and reservoirs on the five rivers where the greatest damage had occurred in 1927—the Lamoille, Missisquoi, Passumpsic, White, and Winooski—but the state rejected this plan. Vermont did, however, strongly support dams in the Winooski River Valley, where the worst damage had occurred. In 1934, with the help of the Civilian Conservation Corps (CCC) and supervised by the Army Corps of Engineers, work began on three dams in the valley: Barre Dam, Wrightsville Dam, and Little River Dam (see map 5.1). The reservoirs behind most of these proposed flood-control dams are usually empty; they are there simply to hold water in the event of a flood. In March 1936, during further formulation of the larger plans, floods again struck the Connecticut River Valley, hitting Massachusetts and Connecticut the hardest. In response, the New England states began work on an interstate compact to deal with the flood-control issues of the Connecticut River Valley. From Vermont's perspective, the problem was that most of the dams would have to be built upriver, in New Hampshire and Vermont, and that these dams would flood some of the states' narrow valleys and their prime farmland. The compact agreed to by the states would feature eight dams, three in Vermont. But the compact was never approved by Congress, due to a battle between the states—led by Vermont Governor Aiken—and the federal government over control of the dams. The states claimed the dams were only for flood control, while Washington, fearful that the states would grant the authority to create electricity at the sites to private utilities, wanted control over power as well as flood control. Congress sided with the administration, passing a law giving the federal government the power to build the dams and control power generated there, regardless of state wishes. The first Vermont project was started at Union Village in the Ompompanoosuc River Valley in 1938, but Aiken and Vermont kept up their opposition and that dam was never built.

A new plan devised by the Army Corps of Engineers called for twenty dams in the Connecticut River Basin, including ten in Vermont. The package called for a dam in West Dummerston that would have flooded more than five thousand acres in the scenic West River Valley. The Vermont delegation was successful in preventing this plan from gaining approval in Congress. The Corps scaled back its plans; nonetheless, there was strong opposition in the late 1940s to proposed flood-control dams at Williamsville (in the West River Valley), North Springfield, Gaysville (in the White River Valley), Cambridgeport, South Tunbridge, and Ludlow. A 1944 proposal to raise the Wilder Dam on the Connecticut River met similar protest, but this project went ahead. In 1947, the Corps issued a new flood-control plan, this time without the dams at Ludlow, Victory, and Williamsville. The following year, Vermont Governor Ernest Gibson asked, "Where would our vision and foresight be if today we converted more of our valleys and ravines into stagnant and unsightly reservoirs and tomorrow we obtained abundant power from a carload of uranium?" In an ironic sense, the governor's vision came true, as Vermont has gone on to become the state that generates the greatest percentage of its electricity via nuclear power.

The flood-control battle finally came to an end in 1951. The four Connecticut River Valley states agreed on a flood-control plan for the entire basin, a plan that featured six Vermont dams: Ball Mountain, Bloomfield, Groton Pond, North Hartland, Victory, and West Townshend. Dams were eventually built by the Corps at Ball Mountain and West Townshend (West River), North Springfield (Black River), North Hartland (Ottauquechee River), and Union Village (Ompompanoosuc River). This entire controversy demonstrates the difficulties of interstate and federal approaches to flood control—and foreshadows those of water-pollution control; it underscores the difficulty of controlling issues outside of a state's borders despite their effects within the state.

The rise of conservation and the establishment of town forests, state forests, and a national forest during the period from the late 1880s through 1950 centrally affected the Vermont landscape. As the forest began to come back because of market-based forces on Vermont agriculture, the conservation movement led to its proactive management. State agencies managed wildlife species that returned, expanded in population, or were reintroduced to the improving habitat in Vermont under new fish-and-game laws. Forest management helped increase the amount of forest through reforestation, increased the quality of lumber produced through forestry, and helped to control disease and fire in the returning woods. The increasing public lands were managed for multiple uses: high-quality timber production, watershed protection, demonstration of forest management to private landowners, recreation (including skiing and tourism, discussed below), and wildlife. The period was also important for what did not happen: construction of the Green Mountain Parkway or a vast resettlement program. The former would have greatly altered the wild nature of the Green Mountains, the latter might have greatly increased the amount of uninhabited forestland in the state. Furthermore, environmental historian Richard Judd argues, the nascent conservation movement of the late nineteenth century in northern New England was crucial in launching and guiding the national movement. "This human-centered rural perspective," he writes, "helped lay the political groundwork for a vigorous conservation movement in the 1890s."

A Slowdown in Human Population Growth

The late nineteenth and early twentieth centuries are often portrayed—from a human perspective, of course—as a period of decline in Vermont. The foundation of this thesis is based on human-population statistics. For instance, 59 percent of Vermont towns lost population between 1870 and 1880, 81 percent lost population between 1880 and 1890, and 72 percent lost population in the final decade of the century. As of 1910, 129 Vermont towns had reached their population peak prior to 1850. In 1930, forty towns were growing steadily while two hundred were declining or stagnant. Another often-cited figure is that in 1880, the number of Vermont natives living outside the state equaled 54 percent of the population of the state, the highest such level for any state (New Hampshire was a distant second at 37 percent).

Overall, Vermont's population increased slightly over 12 percent between 1870 and 1950 (only 8 percent from 1870 to 1940) (see table 5.1). During two decades during this span, the state's population actually declined (1910 to 1920, 1930 to 1940), and in only one decade (1940 to 1950) did population growth exceed 5 percent. These figures are in stark contrast to national population growth over these eighty years. The keys to this slowed growth were, in relative order of importance, decreased immigration, increased emigration, and decreased fertility.

The population declines that occurred between 1850 and 1910 were centered in the Connecticut River Valley counties; populations in the other Vermont regions grew slightly. Throughout the period, Rutland, Chittenden, and Windsor were always among the four most populous counties in the state, with Washington replacing Franklin County in 1900. Rutland County was the most populous until it was overcome by Chittenden in 1940, which has since remained the largest in terms of population. The largest cities were Burlington (the largest every census year but 1880), Rutland (second largest every year but 1880, when it was the largest), Barre (beginning in 1890), Bennington (every year but 1890), Brattleboro, and Saint Albans from 1870 to 1890 and Saint Johnsbury from 1900 to 1920.

Although New England as a whole first became more urban (defined as greater than 50 percent living in towns over 2,500) than rural in 1890, Vermont remained overwhelmingly rural. The state's population was roughly 80 percent rural in 1900, though this figure declined to 67 percent by 1930 as people headed to cities. (This trend more or less halted about this time; Vermont's population remains approximately 67 percent rural today, though it has dipped a bit lower at times.) Growth was concentrated in the largest urban areas. For instance, in 1850 no Vermont town or city had a population over 7,500; by 1910 there were seven such municipalities encompassing over 20 percent of the state's population.

The most complete study of these population trends in Vermont is Hal Barron's study of Chelsea from 1784 to 1900. A hill town thirteen miles from the railroad in the 1850s, Chelsea saw its population stabilize from 1830 to 1850 and then go into decline for the remainder of the nineteenth century due to increased emigration, reduced immigration, and lower natural increase (because of reduced fertility). Indeed, from 1840 to 1900 the town's population declined by over 40 percent. Chelsea was not alone in this decline; over one hundred of Vermont's 238 towns lost population during this period. Seventy towns and cities grew, however, illustrating the more complex nature of population changes in the state. "Rather than viewing these changes as expected characteristics of older agrarian societies," writes Barron, "the historians of the region have long spoken instead in terms of extraordinary decline and decay." The population decline that did occur was not simply because of economic problems but rather was part of the aging of older agrarian communities. "As in so many aspects of American history," Barron further writes, "trends in New England anticipated analogous development in other regions, and rural New England became the first agricultural area to grow old."

Barron's study of Chelsea—of those who stayed behind—revealed a town not in wholesale decline but adjusting to a new equilibrium. The number, size, location, and output of farms in the town remained roughly the same: "The constant number of farms and level of farm production suggest that agricultural development in Chelsea had reached the limits of its growth and had leveled off." Outside of agriculture, change was taking place but not an overall decline. Small factories and artisans declined in response to increased competition arriving via rail, yet retail merchants and professionals remained relatively constant. Those who remained in Chelsea had strong connections to the town, such as owning or expecting to own land, and they were less likely to leave. Hence, this selective emigration brought a kind of stability to the town. Two other studies support Barron's findings: One of seven Vermont towns and one of Jericho both indicate that it was not economic decline so much as smaller farm families and larger farms that led to population declines.

These population trends left Vermont the most homogeneous state in the country in 1930: seventy-two percent of those living in the state had been born there. Although most commentators saw these trends as indicative of decline, at least one Vermont historian had a different interpretation. "What this seems to show, more than anything else," wrote Edmund Fuller in 1952,

> is that Vermont quite naturally adapted itself to this "optimum" [population level] —that is to say, the best natural condition for the state's welfare. . . . A hundred years or more shows that the state can accommodate, very comfortably, at least that many people. It seems like common sense to think that if it would be unhealthy for the state to fall heavily below such a figure, it might be equally unhealthy to have the population forced too far above that level.

With the state's population now approaching 600,000, it would be interesting to hear what Fuller would have to say today.

Admiring Scenery, Skiing Down Mountains: The Rise of Tourism

A great new force on the landscape—tourism and recreation—really began to make its presence felt in the post–Civil War period. Prior to this, tourism in Vermont revolved around mineral springs, especially those at Clarendon Springs (the first significant one, established in the 1770s), Highgate Springs, Sheldon Springs, Middletown Springs, and Brattleboro. The arrival of the railroads gave a further boost to the mineral springs; by the end of the nineteenth century there were over 130 such springs in the state, more than thirty of which featured hotels. Despite the railroads, the era of the mineral-springs resort began to decline during the Civil War due to lagging interest in water cures, the loss of southern guests, and the rise of a new kind of tourism.

The focal point of aesthetic tourism in the period following the Civil War was a sublime and picturesque landscape, a trend built upon the influence of the Hudson River school of painters, the Romantic literary tradition, and the Transcendentalists. Vermont was not a major tourist destination. Rather, in the Northeast the state lost out to the more sublime Adirondacks and White Mountains, with their craggy peaks, and the rocky Maine coast. Vermont's Green Mountains were dismissed by some as green hills. Nonetheless, many tourists did come to visit Vermont's scenic features, and large resorts developed nearby. Indeed, as early as 1845, the state geologist advocated promoting Vermont's scenery to tourists. In 1858, the first summit house was opened at the top of Mount Mansfield, and twelve years later a road up the mountain was completed. Summit houses were also built on

Bread Loaf, Camel's Hump, Killington Peak, Mount Anthony, Mount Ascutney, and Snake Mountain, although these all folded by 1900. Other major inns were in Manchester and Woodstock.

Beginning in the 1880s, though, efforts began to celebrate and sell Vermont's pastoral landscape. During this decade, for instance, *Harper's* magazine featured an article on Vermont that highlighted its pastoral qualities, the Vermont Central Railroad began to focus on these characteristics in its travel brochures, and photography of the state concentrated on the bucolic rather than the mountainous. The state began selling Vermont to tourists in the late nineteenth century, largely in response to the decline in rural population and number of farms, especially hill farms, and in order to woo some of the tourists heading to neighboring states. With the railroads in the state, the Board of Agriculture developed advertising to promote Vermont as a place for urban residents to get away from urban problems and to recover rural values such as virtue and simplicity, as a place to rediscover one's past. These advertising campaigns appealed both to middle-class people seeking less expensive, closer-to-home vacations, who would often come to the state as farm boarders—the most important source of tourism in the early years—and to the wealthier, who could purchase an abandoned farm as a vacation house. In 1893, for instance, the Board of Agriculture published the *List of Desirable Vermont Farms*. Come to Vermont and escape "the heat, the dust, and disease of the cities and become strong by close communion with nature, surrounded with her richest privileges" read a state brochure from the time. The state became even more serious about tourism promotion in 1911, creating a Bureau of Publicity, the first state to do so. Its first publication: *Vermont, Designed by the Creator for the Playground of the Continent*. This state action culminated for this period in 1946 when the Vermont Development Commission founded—and funded—*Vermont Life* magazine to promote tourism and Vermont generally.

As the twentieth century progressed, the automobile further stimulated tourism in Vermont. Eventually, this auto-based recreation spurred the development of gas stations, motels, and restaurants—a service industry. In the middle 1920s, half the cars on Vermont highways in the summer were from out of state. By the end of that decade, more than one million summer tourists came to the state. In the middle 1940s, nearly 7 percent of Vermont houses were seasonal. Other popular tourist activities drawing out-of-state visitors were hunting and fishing, horseback riding, and summer camps for boys and girls.

Without question, though, the major recreational catalyst to tourism in Vermont involved winter sports. This dimension began to develop in the 1920s and 1930s. Stowe was the early focal point of skiing in Vermont. Winter carnivals began there in 1921, and soon people were skiing down the unplowed summit road. The Stowe Ski Club formed in 1931, and trails were cut on the mountain over the next years, some by the CCC. A ski school opened in 1934 and a rope tow in 1935. Ski trains were running to Waterbury the following year. In 1940, the Mount Mansfield Company leased some state forest, and by the next year the longest and highest chairlift in the world was built halfway up the mountain.

Development also took place at a number of other locations throughout the state. Brattleboro featured a ski jump in 1922; the first ski tow in the United States began operation in Woodstock in 1934, the first T-bar in 1940. The CCC helped skiing expand in the state by building trails at Mount Ascutney (on state park lands), Burke Mountain, Jay Peak, Killington, and Okemo Mountain (on state forest lands), though it would be many years before large-scale ski areas were developed on these mountains. Other areas that began in the 1930s included Bromley (on national and state forest lands), Hogback in Marlboro (on state park lands), Pico, and Suicide Six. The state actively supported the development of ski areas, first by establishing state forests in promising ski terrain and helping new areas begin, and later by making special appropriations to build access roads to ski areas such as at Mount Mansfield, Killington, and Mount Snow. In 1948, the state had fifty-five ski areas. With skiing, Vermont became a tourist destination in the winter as well as the summer. To help stimulate business in the fall, the state started to promote the fall foliage season, making Vermont a three-season destination. (Spring in Vermont goes unaffectionately under the label "mud season.")

Another recreational component of this tourism surge was hiking. About forty Vermont mountains had known hiking trails by the beginning of the twentieth century. Building on this interest, the Green Mountain Club, established in 1910, built the Long Trail in sections over the next twenty years. This "footpath in the wilderness"—the oldest long-distance trail in the United States—is roughly 265 miles long, running along the spine of the Green Mountains from Massachusetts to Quebec. With seventy shelters along the way, the Long Trail drew (and continues to draw) many Vermonters and out-of-state visitors into the mountains of Vermont. The Appalachian Trail (AT), which runs from Georgia to Maine, coincides with the Long Trail from the Massachusetts border to Sherburne Pass, where the AT heads east into New Hampshire. Completed seven years after the Long Trail, over 130 miles of the AT run through Vermont.

Despite the positive economic effects of tourism in Vermont as a whole, not all parts of the state enjoyed these benefits equally. Furthermore, Vermont lagged behind the other New England states as a tourist destination, receiving less income from seasonal dwellings than any other state and ranking fifth (as of 1950) in overall income from vacationers.

The foundations of Vermont's current tourism economy were laid during the first half of the twentieth century. The focus was on a pastoral landscape, not a wild one. People came to view and enjoy a landscape of small farms and managed forests. This translated into the widespread development of roads, trails, and services for guests, even in the highest, most remote mountains. The managed landscape at the heart of Vermont's nascent tourist industry meshed well with the goals of conservation discussed previously. These connections could be seen empirically in a number of proto-environmental initiatives launched by the tourist industry to protect the state's scenic beauty. For instance, the Vermont Association for Billboard Restriction, working with the State Development Commission, had moderate success in removing large billboards and controlling smaller ones. In addition, the secretary of the Vermont Chamber of Commerce led a campaign to clean the state's rivers and streams.

Mining, Industry, and Infrastructure

The four mining industries discussed in chapter 4—copper, marble, granite, and slate—each reached their peaks of production during the late nineteenth and early twentieth centuries, and an important new mining venture—talc, asbestos, and soapstone—was established in the early 1900s. Copper production in Vermont first peaked in 1880 at 5.5 million pounds of metallic copper (compared to the 20 million pounds produced in the Lake Superior region). Production then steadily declined to less than a quarter million pounds in 1906. In order of productivity, Vermont's copper-mining towns were Strafford, Vershire, and Corinth. The railroad greatly aided the Vermont copper industry, allowing the ore to be shipped to smelters in Boston and elsewhere. Vermont also had its own smelter, beginning in the 1860s, the emissions of which led to very acidic soils in the region.

The deposits were worked on and off until the World War II–induced copper shortage led to the reopening of the mines at full production in 1943. With much better mining techniques, production reached 6 million pounds in 1946 and 1949. The Korean War further stimulated production, leading to peaks of 8.5 million pounds in 1954 and 1955. By 1958, though, the richest ores had been depleted, and new sources of copper were being developed elsewhere. Consequently, the Vermont copper mines closed down, apparently for good. A visitor to the mines in Vershire finds an overgrown and abandoned landscape, only hinting at the major activity of forty years ago. Overall, the Vermont mines produced about 145 million pounds of copper, over two thirds from the Strafford mines and mostly following the war boom of the early 1940s. The acid drainage from these mines has had and continues to have a significant effect on the nearby aquatic ecosystems. For instance, as of the mid-1980s, there were no trout in the Ompompanoosuc River between West Fairlee and Thetford or in the West Branch of the river below South Strafford.

From 1860 through 1900, the Vermont marble industry grew tremendously, with the value of marble sold rising from $500,000 to $3.6 million. Railroads to transport this heavy stone to out-of-state markets were central to this growth. In addition, concentration accompanied this growth, as the Vermont Marble Company acquired other producers, dominating production in the state by 1910. The number of quarries declined with this concentration, to roughly twenty active ones in 1915. Around this time, growth slowed due to competition with granite and marble from elsewhere. Nonetheless, Vermont was the leading marble producer in the United States between 1880 and 1930, when it was surpassed by Tennessee.

Granite was similarly important in Vermont. Barre granite quarries took off in 1888, when a rail spur to them was built. Barre then became the granite capital of the United States. The railroads helped the marketing of the granite throughout the country and also simplified getting the stone from the quarries to the cutting sheds. By 1914, granite overtook marble in importance in Vermont. Although some granite, like marble, was used in constructing public buildings, as early as 1915 three quarters of the granite sold was for memorials. At the industry's peak in the early 1900s, there were nearly eighty granite quarries in Vermont, over half in Barre and another batch concentrated in Woodbury, with the rest scattered throughout the state. Throughout this period, Vermont was first or second in granite production in the country. The monetary peak for the industry came in 1945 with the high demand for war memorials.

As was the case for the other stones, slate production in Vermont was greatly stimulated by the coming of the railroad. By the mid-1880s, over seventy firms were quarrying slate in west-central Vermont, primarily for roofing. In terms of value of production, the slate industry reached its peak around 1920. After this time, people and construction companies began to use cheaper materials for roofs. From 1870 to 1950, Vermont consistently ranked second to Pennsylvania in terms of slate production.

Mining for talc, asbestos, and soapstone began in the early 1900s. The three minerals are related geologically and are found in alternating bands on the east side of the Green Mountains from the Canadian border all the way to Windham County. The major talc mines were in Johnson, Rochester, and Windham, and all relied on good rail connections to move the material. By the middle teens, Vermont was the second leading talc producer in the country. Asbestos mining focused on Belvidere Mountain in Eden and Lowell, near the Canadian border. It began in the early 1900s but essentially stopped until the 1930s. At that time, the Vermont mine produced over 80 percent of the asbestos in the country (though the vast majority of the mineral used in the United States was imported from Quebec). Production was strong through World War II and into the 1950s. Soapstone, quarried in a number of towns, has been less important than these other minerals in terms of amounts produced and the value of that production.

In terms of landscape disturbance, the most important mining in Vermont has been for sand and gravel. With the rising use of concrete for construction and gravel for unpaved roads, such quarrying greatly increased. This mining is the most widespread and noticeable, as firms typically take these materials from glacial deposits, especially along the western slope of the Green Mountains (as described in chapter 2). Two minor mineral industries that peaked during this period were lime, from the state's limestone deposits, and fine-quality kaolin clay produced from beds in Monkton, beginning in 1792. Production of both stopped around 1970.

Throughout the period from 1870 to 1950, Vermont remained an industrial backwater, with textile production the most significant industry in the state. The number of textile mills declined after the stimulus of the Civil War, but production remained strong, peaking around 1920. Even at this time, though, Vermont's production was only about 5 percent of the New England total. The industry declined overall following this peak, though major mills remained in the Black River Valley, Burlington, and Winooski until around 1950. The two other major industries in Vermont were machine-tool manufacturing in the Springfield-Windsor area and the making of platform scales in Saint Johnsbury and Rutland. Indeed, roughly three quarters of the nation's scales were produced in Vermont in 1950. Furthermore, better transportation exposed local small-scale manufacturing to fierce competition from beyond, causing many factories to fold, which in turn brought further decline to the smaller towns (see chapter 4). Even with this limited industry in Vermont, historian Edmund Fuller was already expressing concerns over out-of-state control of Vermont industries, underscoring in yet another way how Vermonters were losing control over their lives—and over the landscape of the state. He wondered if such outside ownership "contains any hidden danger of losing control over our own economic destinies; of giving rise to a captive economy in which decisions made in New York City, or Detroit, could determine the welfare of Vermont communities." These are questions Vermonters continue to wrestle with today.

The lack of significant industrial development and large urban centers spared Vermont's waterways from the frightful water pollution—human wastes, acids, dyes, sawdust—and extensive engineering to control the water for power sources that greatly affected the rivers of southern New England. By the middle of the nineteenth century, the fisheries of the Merrimack

River in Massachusetts and New Hampshire, for instance, were destroyed. Vermont did not entirely escape these major changes to aquatic ecosystems, though. Dams and mills altered the temperature, speed, volume, and composition of the water, causing significant ecological changes. They also blocked migratory fish runs and changed river flows. As early as 1870, all the New England states except Rhode Island had fish commissions in place to help restore migrating fish to the states' waters through restocking (especially salmon and shad in the Connecticut River watershed) and building fishways.

Railroads remained the major mode of transportation into the twentieth century, but very little new track was laid after the 1870s. Sixty years later, many of the spur lines were closed because of competition from cars and trucks. Roads rose in importance as the railroad era entered its decline and more tourists started to come to Vermont. A gas tax was first established in 1892, with the state distributing funds to towns to improve roads. Vermont was one of the first states to establish such a program, partially to help move fluid milk throughout the state. To bring some central control to the road system in Vermont, a State Highway Commissioner was appointed in 1898. In response to the 1927 flood, the state government took on increased responsibilities for road and bridge maintenance since the towns could not afford to pay for the necessary repairs. The flood was also the catalyst to create a state highway system—the first time the state government would have primary responsibility for the major roads—and to build more hard-surface roads. Furthermore, the flood accelerated the shift away from the railroad toward the car and truck since many of the rail lines were not restored as the state poured money into highways.

The number of cars in Vermont tripled in the 1920s to 90,000. More cars led to demands for better-quality roads, and by 1943 10 percent of the state's roads were paved. Vermont's roads, however, have always proved problematic due to the state's low population and rural nature as well as the hilly and mountainous topography and freeze-thaw cycles throughout the state. The shift to private motor vehicles as the primary mode of transportation left Vermont again in a transportation backwater. Prior to the interstate highways, the drive from Burlington to Boston could take upward of seven hours on winding two-lane roads. The state fell back into a kind of economic isolationism similar to that before the railroads were built, as businesses were reluctant to locate in a place so difficult to reach.

In terms of mining, industry, and transportation systems, only the latter had a widespread effect on the Vermont landscape. Though the state was home to significant mining industries, they disturbed relatively little of the state's landscape—unlike the coal mining of Pennsylvania or Kentucky. The Industrial Revolution and its negative effects on the air and water largely bypassed Vermont because of its relative isolation from major manufacturing and urban centers. Industry's biggest effect on the state was not pollution but mills to create power, which altered river and stream ecology throughout Vermont. The rail system affected the landscape by concentrating development and introducing Vermont products to larger markets, a process that hurt many Vermont industries (such as textiles) but favored others (such as dairy). The spreading road system helped develop the tourism that was to rapidly assume center stage in influencing the way people thought about the Vermont landscape but also helped put the state at further economic disadvantage, at least through 1950.

The changes during this short century of decline and recovery were the result of both active and passive forces, both internal and external. The decline of general farming and the rise of specialized dairy farming were primarily due to external market pressures. These changes in agriculture were passive forces in the return of the forest, which was aided by the active conservation movement. Furthermore, these changes combined with a new tourism industry that sought to develop and protect a certain kind of Vermont landscape. The new themes guiding change from 1870 through 1950—dairy farming, a returning forest, conservation, tourism—remain the dominant ones in the modern Vermont landscape.

6

Creating the Modern Vermont Landscape, 1950–1995

THE VERMONT LANDSCAPE of the last five decades has been affected primarily by two sets of forces. The first set was quite different from those that had been affecting the state since the middle of the nineteenth century: population growth and economic development. The state's population increased by more than half, accompanied by economic development—especially tourism—and the interstate highway system. The second set of forces was largely a continuation of trends begun in the mid-nineteenth or early twentieth century: the decline of agriculture, the return of forests, and the continuation of conservation. This mix of new and old forces, as well as geological and biological forces, combined to create today's Vermont.

Population Growth, Interstate Highways, Development Pressures, and Act 250

The major force shaping the current Vermont landscape was human population growth. From 1960 to 1990, Vermont added nearly 173,000 permanent residents—almost a 50 percent increase. This was the greatest number of new humans ever to come to Vermont over a thirty-year span, surpassing even the period of the republic and early statehood. This growth was due to several factors and led to a number of fundamental changes in the state. The main causes for this growth were the improved highway system (especially the interstate highways), the continued growth of tourism (which brought many to Vermont who decided to stay), and the changing nature of the economy toward services and light manufacturing (sectors in which Vermont could be quite competitive). The greatest consequences of this growth were dramatic population increases in some communities—especially Chittenden County and towns near ski areas—tremendous land-development pressures, and various attempts to control development, most notably Act 250.

The state's population began to grow rapidly after 1960 (see table 6.1). For the first time since the early 1800s, substantially more people were moving to Vermont than were moving out of the state. Immigrants accounted for nearly a third of Vermont's growth in the 1960s, and nearly 60 percent in the 1970s. In 1960, 72 percent of Vermonters were born in the state; in 1980, 60 percent were (including children of newly arrived Vermonters). Such growth was occurring throughout northern New England, as Maine, New Hampshire, and Vermont all grew at over 10 percent rates in the 1970s while none of the southern New England states grew more than 3 percent. A final indicator of this growth was that no native Vermonter has been elected governor since 1970, paralleling the first four decades of statehood, the other period of rapid population growth (when the first native Vermonter was not elected governor until 1835).

In each of the three decades from 1960 to 1990, Vermont had double-digit population growth—peaking at 15 percent in the 1970s, and, for the first time since the early nineteenth century, the state grew faster than the United States as a whole. From 1850 through 1960, Vermont's population was between 300,000 and 400,000. By 1995, it was approaching 600,000. Although Vermont had the second smallest population of any state, the numbers still indicated a tremendous surge in human population. The state's growth finally slowed in the 1990s, to less than 4 percent from 1990 to 1995. Although this rate was below the national growth rate, Vermont had surpassed New Hampshire to become the fastest-growing state in New England.

There was no clear pattern to the growth: Bennington, Chittenden, Grand Isle, Lamoille, and Orange Counties, scattered throughout Vermont, all grew faster than the state average from 1960 to 1980 (see map 6.1). Most notable was the growth of Chittenden County during this period. Already the state's largest county by nearly a factor of two in 1960, it grew an aston-

TABLE 6.1.
Human and Animal Populations in Vermont, 1960–1995

Year	Human population	Density (sq. mile)	Sheep population	Cow population
1960	389,881	42.0	12,000	240,928
1970	444,330	47.9	6,715	195,828
1980	511,272	55.1	8,200	200,000
1990	562,758	60.7	29,500	181,000
1995	585,000	63.1	17,000	172,000

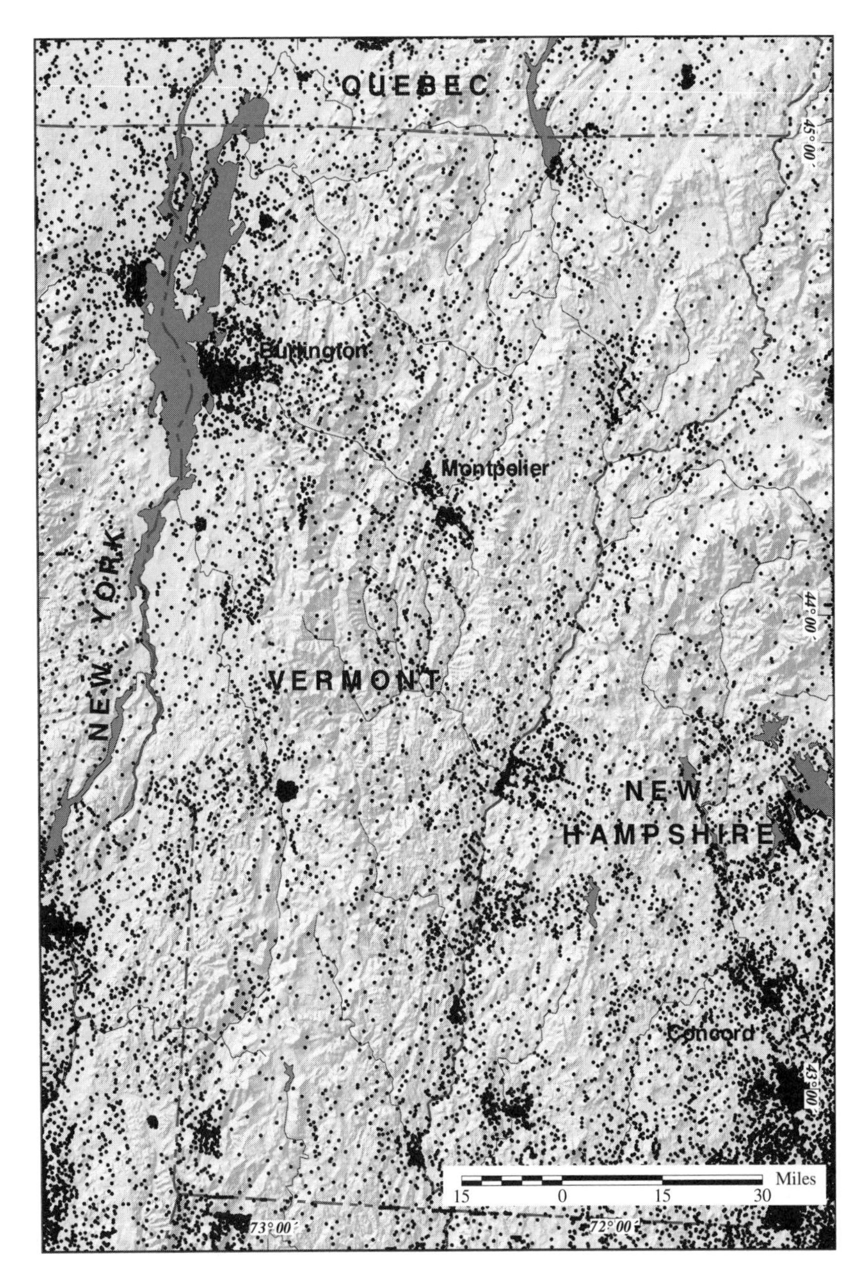

MAP 6.1. Greater Vermont population density, 1990. Each dot (•) represents one hundred people.

ishing 77 percent over the next three decades, becoming a real urban center in a state that had long lacked one. Although Burlington and Rutland remained the state's two largest cities, by 1990 three Burlington suburbs—Essex, Colchester, and South Burlington—were among the six largest municipalities in the state. Nonetheless, the United States Census Bureau had to change its rules for Burlington to count as a metropolitan area. Prior to 1980, such an area needed an urban center of 50,000, a population level Burlington still did not meet. In 1980, this requirement was waived, and the Burlington Metropolitan Statistical Area was created, with a population of 115,000.

Despite this increased growth in Chittenden County, Vermont still remained overwhelmingly rural. Indeed, the percent of people living in rural areas increased from 62 percent in 1960 to nearly 68 percent in 1990. Furthermore, Vermont was far and away the most rural state in the nation (with Maine third and New Hampshire seventh). Compared to the southern New England states, Vermont remained sparsely populated: in 1990, Vermont's population density was nearly 61 persons per square mile, Connecticut's was 678, and Massachusetts's was 768. Negative aspects of this high rural population included the increasing conversion of villages and towns into bedroom communities for larger towns. This led to, among other things, more commuting and the building of more roads and houses in forests and fields, fragmenting the landscape for agriculture, forestry, and wildlife. Commuting was also stimulated by high property values in certain towns (such as Stowe), forcing many working in the town to commute from places where they could afford to live (such as Morrisville).

Along with these migrants, another group of people made its presence felt during this period: the Abenaki, the native people living in Vermont when the Europeans arrived. Throughout the 1800s and much of the 1900s, the Abenaki were an unmentioned people in Vermont. Histories from the period reported that no Indians had lived in Vermont when Europeans came or that they had all been killed off or moved on by 1800. Ethnic pride fueled an Abenaki resurgence, demonstrated by an increase in the number of Vermonters identifying themselves as Native American from 229 in 1970 to 1,700 in 1975. From the late 1700s through the 1970s, those Abenaki who had remained in Vermont lived as transients; in settlements in Swanton, Saint Albans, and Grand Isle; or in areas so remote that they were rarely seen by non-Indians. Many lived from the land—hunting, fishing, trapping, gardening, herb collecting—as much as possible. The mid-1990s Abenaki population in the state was approximately two thousand, centered in the Swanton-Highgate-Saint Albans area. They were engaged in legal battles with the state government over hunting and fishing rights and with the federal government over legal recognition of the tribe.

A number of commentators point to the interstate highways as the crucial catalyst for the increased population growth and accompanying changes to the Vermont landscape. "The most direct cause of this [population] increase," writes geographer Harold Meeks, "was the approaching completion of the National Defense Highways [the interstates]." Vermont author Joe Sherman writes that "the interstates and the new ski areas they served changed the face of Vermont more than anything had since the Ice Age." The interstate-highway system (formally named the National System of Interstate and Defense Highways) was authorized by Congress in the mid-1950s. Ostensibly part of the cold-war national defense system—to move troops and help move civilians in a war—as well as a transportation system, these highways became the largest public-works project in the nation's history and have had profound effects on the economy and on residential patterns, as well as the ecology, throughout the United States. Vermont has more than three hundred miles of interstate (see map 6.2). These roads, completed by the mid-1970s, fully integrate Vermont into a high-grade regional-highway system. Prior to these interstates, Vermont's highway system was not on par with those found throughout the nation; hard-surface roads were still uncommon in 1950. This meant tourists were hesitant to come to Vermont, as were businesses that were growing increasingly dependent on highway connections.

Some critics argue that the interstates were substantially overbuilt in Vermont. In 1970, for instance, Vermont per capita interstate mileage was three and a half times the national average though traffic volume was one third the national average. Nonetheless, the state also sought to upgrade the rest of its highways in order to make better use of the interstates. This road-building craze came to an end in the 1970s when the state decided against building a major four-lane east-west highway in the state and when a plan to make Route 7 in western Vermont four lanes for its entire length was derailed.

The obvious effect of these interstates was that they made it easier for people to come to Vermont and for Vermont businesses to ship products elsewhere. The shorter travel time to Vermont made it a more attractive place for tourism, especially long weekends for skiing, viewing the fall foliage, and just getting away. The ski areas of southern Vermont, for instance, were now just two and a half hours from Boston and three hours from New York City. Eventually, many of those who came as tourists decided they wanted a second home or condominium, which fueled the development boom.

The improved roads and the changing economy also made Vermont more attractive as a place to do business than at any previous time in its history. The better roads placed Vermont in the heart of a dynamic region between Boston, New York City, and Montreal. From Burlington to Bos-

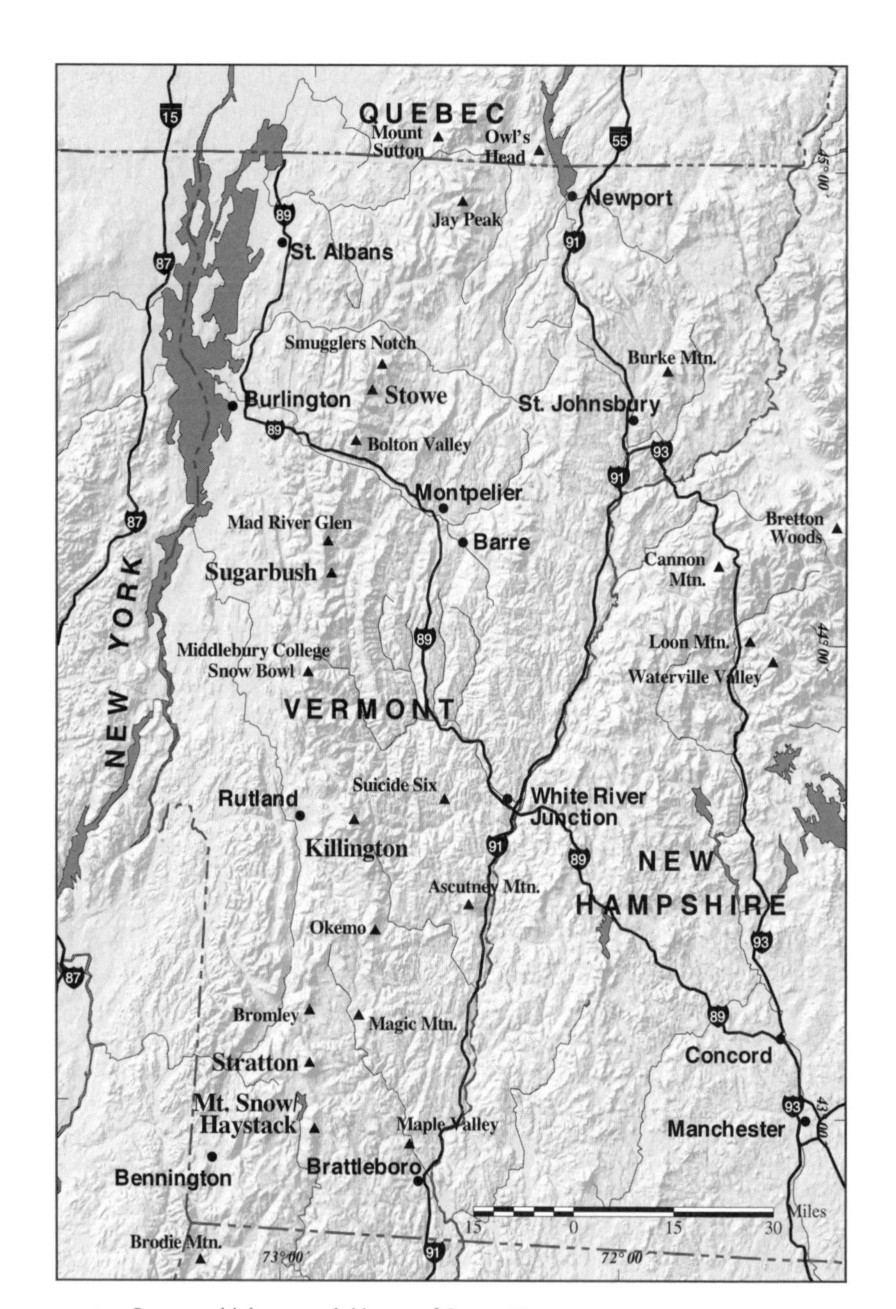

MAP 6.2. Interstate highways and ski areas of Greater Vermont, 1995.

ton was now less than a four-hour drive. As the national economy began to emphasize services and light manufacturing, products less affected by high transportation costs, firms looked at different factors in deciding where to locate. For some firms such as IBM, Vermont's quality of life became a major attraction.

Another change altered the political landscape of Vermont. The United States Supreme Court's *Baker v. Carr* decision in 1962 dramatically reduced the power of small towns in the state. The ruling required reapportioning state legislatures so that each person's vote was equivalent to every other person's vote. Prior to this decision, each town and city had one representative in the state House. This earlier system had greatly increased the political power of small towns and rural areas; in the House, a town with a population of several hundred had the same vote—one—as did Burlington with 35,000 people. Under the new system, the state's urban centers gained power at the expense of the rural areas, providing for a political system more responsive to demographic and social changes.

It was not so much the new residents as the temporary ones that served as the catalyst for Act 250, Vermont's land-use regulatory law. Second-home developments near ski areas in southern Vermont were overwhelming local town resources. In addition, some were built on steep slopes, leading to erosion and water-pollution problems. Governor Deane Davis appointed the Commission on Environmental Control (known as the Gibb Commission, chaired by Senator Arthur Gibb) in 1969 to study the issue and make recommendations. Davis said at the time that Vermont was in trouble "unless we did something about the invasion of the state by people and the type of quick development that was going on to make a fast buck." Many of the recommendations were passed as Act 250 the following year. This wide-ranging and progressive piece of legislation had two main parts: a permit process and a state land-use plan. The developer is required to receive an Act 250 permit from the relevant district environmental commission (there are nine districts in the state as well as a statewide Environmental Board) before subdividing land for sale in lots less than ten acres in size, or for the following types of development: a subdivision of ten or more units, any commercial or industrial uses of more than ten acres (more than one acre in towns with no zoning), any state or municipal project of more than ten acres, or any development other than farming or logging on land above 2,500 feet. There are ten criteria on which the project is judged, including air and water pollution, water supply, soil erosion, traffic, municipal services (including schools), scenic beauty, wildlife habitat, and conformance with existing plans. Farming and forestry operations are exempted from the law. These criteria were amended in 1973 to help protect shorelines, floodplains, and prime farmland and forestland from development.

The second component of Act 250 directed the state to develop three plans. The Interim Land Capability Plan was adopted by the legislature in 1972, and the Land Capability and Development Plan was approved in 1973. This second plan included additional permit criteria and clarification of the existing criteria. The third plan, the Land Use Plan, indicated where growth was to be concentrated and where it was to be limited. Lands in the state were classified as urban, village, rural, natural resource, conservation, shoreline, and roadside. Future development was to be encouraged in urban, village, and rural areas. The natural-resource areas—roughly half the state—could only have one primary dwelling per twenty-five acres. Development in the conservation areas, roughly one third of Vermont, was to be even more restricted—no more than one principle dwelling per hundred acres—and was to be under the control of the state government. The plan met with immediate opposition, as it would restrict the potential development of virtually all of the land in the state, not just big developments built by out of staters, making many property owners angry. The move away from town control also alienated many Vermonters. Finally, the state's economy had stalled, so the threat of rampant development seemed much more remote. For all of these reasons, the legislature refused to even vote on the Land Use Plan in 1974, and a revised version failed the following year. In 1983, this part of Act 250 was repealed. This meant that only half of Act 250 was in place. The permit process designed to control specific projects had worked, but the state plan to guide where growth would occur was absent.

The great success of Act 250 has been to control and improve the quality of development in the state. The law has not stopped development; the vast majority of permits requested were granted, though most of these included provisions designed to address concerns that arose in the review process. In the 1970s, for instance, over 3,500 applications for an Act 250 permit were filed, yet only one hundred were rejected outright. Perhaps the most notable was the proposal for a mall in Williston, fought by Williston residents and downtown Burlington businesses. Ironically, despite the permit denial, the site was not protected from development. Rather, it has seen massive development—including new branches of Wal-Mart, Home Depot, and Toys "R" Us. An example of a project significantly scaled back was the Haystack Mountain development in Wilmington, which was scaled down from over two thousand houses to fewer than one hundred due to water-pollution concerns.

Despite the overall success of Act 250, there were problems with the law as well. Most notably, it did not deal with cumulative development, so strip development continued unabated; this was most pervasive around urban centers such as Burlington and Rutland. In addition, it spawned more scattered and piecemeal development, which some developers used to avoid the permit process. The main reason for these problems was the lack of the Land Use Plan. Without it, Act 250 was denied much of its potential to shape patterns of development throughout the state, leading to increased fragmentation of the landscape.

Another surge of economic development underscored the shortcomings of Act 250. The legislature responded in 1988 by passing Act 200, a companion piece of legislation meant to encourage town planning. As envisioned, the town plans (working from below) would combine with statewide granting of permits (working from above, through Act 250) to create comprehensive and coherent planning and development. In 1985, 200 of Vermont's 246 towns had planning regulations, and 190 had zoning regulations. Act 200 sought to entice all towns to develop consistent town plans. To be approved by the state, these plans must be consistent with state goals and be compatible with regional and neighboring town plans. The primary incentives for participation in Act 200 were state funds for planning and increased town-planning and regulatory authority if the town plan was approved by the state. A significant backlash by property-rights advocates began soon after the law passed. The result was that over one hundred towns, representing roughly one quarter of the state's population, passed resolutions against participating in the Act 200 planning process. Since then, however, all but a handful of towns participated in the planning process, and by 1997 nearly half the towns in the state had approved plans.

A parallel component to Acts 250 and 200 in efforts to control development was tax policy. In 1973, the legislature passed a special capital-gains tax on profits from land sales. Profits gained on land held for less than six years (unless it was the principal dwelling of a Vermonter) was taxed on a sliding scale, based on the size of the profit and how long the land was held. The highest rate, for land held less than a year, was 60 percent. Five years later, the current-use taxation program began. Responding to dramatically increasing property taxes due to increased land values for development purposes, this program was designed to reduce the tax burden on farmland and forestland by taxing it based on its current use rather than its market value. To enroll in this program, landowners must meet certain qualifications regarding income from farming or forest-management plans and commit to not developing the land. If it is developed, a penalty is paid. Local governments were reimbursed by the state for lost revenues, though the program was rarely fully funded, and in 1996 the legislature stopped the payments, meaning significant tax increases in many towns. The program was viewed as problematic by farmers, forestland owners, conservationists, and government officials because of the political difficulties in financing it. Nonetheless, in 1997 over 1.4 million acres were enrolled in the program.

Property taxes in general played a key role in shaping the Vermont landscape. Such taxes had a long history in Vermont—the first one was levied in 1780, with state land taxes of a penny an acre following in 1791, 1797, 1807, and 1812. Vermonters started differential treatment of outsiders early in their history as well. In 1785, a tax on unimproved lands owned by absentees for speculative purposes was established. If these taxes were not paid within thirty days, the land was sold—usually to Vermonters at especially cheap prices. More recently, increased development pressures led to increased land prices and increased property taxes. This led many farmers and forest owners to subdivide lands or engage in forest cutting in order to pay these taxes, despite the current-use program. Further changes in property taxes flowed from the 1997 case *Brigham v. State*, in which the Vermont Supreme Court ruled that the property tax–based funding system for public education was unconstitutional.

Joe Sherman contends that this focus on development has left Vermonters with a paradox: "Growth threatened Vermont's special sense of place, yet maintaining that sense of place, with its pastoral look, open spaces, and small towns, threatened to make Vermont elitist, an upscale getaway for the rich." This paradox was expressed in another way in 1993 when the National Register of Historic Places placed the entire state of Vermont on its list of most endangered historic sites. The reason: the loss of its rural and village landscapes due to the onslaught of suburban sprawl and strip development.

Since 1950, much about the state of Vermont fundamentally changed. For a state that had seen little growth over the previous hundred years, the changes were incredible. In spite of the surge in human numbers, the Vermont landscape retained much of its rural character (with the exception of Chittenden County). Despite the interstate highways and the changing economy, Vermonters wanted to maintain a pastoral landscape of farms, forests, and villages. This landscape was one that appealed to longtime Vermonters, was one that attracted newcomers, and was central to the state's economic base—tourism and a reputation for quality products and quality of life.

From Forestland to Farmland to Funland: Tourism and the Vermont Landscape

After 1950, Vermont's image became even more important to its economy and way of life. The pastoral image of farms mixed with forests, of small villages with their white houses and churches, represented a simpler and slower version of life in America. The traffic jams, the sprawling suburbs, the violent crime, the blighted environment—these problems did not seem to exist when one crossed into Vermont. This landscape connected with the deep-seated American love and respect for agriculture and small-town life, even though the vast majority of Americans do not seem to want to live such a life. Such an attitude certainly helped support an active conservation and land-protection program, which we will discuss in a later section.

Downhill skiing remained the foundation of tourism in Vermont, especially in the 1950s and 1960s. New ski areas were founded into the 1960s, with the state continuing to aid in their development. In 1946, the state purchased three thousand acres of land on Killington Peak to help stimulate the development of a ski area there, even though the area was five miles from the nearest road. The state also paid for building the necessary access road as well. Killington went on to become the largest ski area in the East, connecting six mountains. The resort acquired neighboring Pico Peak in the mid-1990s and will connect to that mountain as well. The Jay Peak State Forest was acquired in the mid-1950s, and a small ski area was created. Soon after, the area was purchased and further developed by the forest-products company Weyerhaeuser, which owned the land surrounding the state forest. There was also significant growth near the previously existing major ski areas. For instance, the populations of Fayston, Waitsfield, and Warren—the towns where Mad River Glen and Sugarbush are located—more than doubled between 1960 and 1980.

Skiing remained the single most important type of tourism in terms of number of visitors and the amount of money they spend. Vermont primarily served as a source of skiing for the large metropolitan areas in the Northeast. In the mid-1980s, for instance, 20 percent of skiers came from New York, 20 percent from Vermont, 15 percent from Massachusetts, 14 percent from Connecticut, and 12 percent from New Jersey. The state could not compete with the resorts of the western United States and their bigger mountains and greater snowfalls, though, and there has been a shakeout in the ski industry since 1975 as competition and costs have increased. Smaller areas that were unable to attract out-of-state skiers or that could not afford to put in significant snowmaking machinery went out of business, such as Sonnenberg in Barnard. A number of other areas—such as Burke Mountain and Ascutney Mountain—recently returned from bankruptcy, and it is likely that a few of these smaller areas will close in the next ten to twenty years. In 1995, the major areas were—from north to south—Stowe, Sugarbush, Killington, Stratton, and Mount Snow/Haystack. Additional areas included Jay Peak, Burke Mountain, Smugglers Notch, Bolton Valley, Mad River Glen, Middlebury College Snow Bowl, Suicide Six, Okemo, Ascutney Mountain, Magic Mountain, Bromley, and Maple Valley (see map 6.2).

In the 1980s, the attitudes of some Vermonters toward skiing began to change. What had been viewed as a perfect match for Vermont—development that complemented the beauty of the Vermont countryside and mountains—was being viewed as a source of problems. The large ski areas seemed to constantly be seeking to expand their trails, often into remote wildlife habitat, and to expand their snow making, diverting water from streams, which might hurt fish populations. These resorts also stimulated sprawl development and high land prices in small mountain towns. The honeymoon between Vermont and skiing was clearly over. Though it remained a popular and important economic sector, the ski industry must often fight long political battles to expand (as Killington has done) or to increase water withdrawal for snow making (as Sugarbush has done). This was a far cry from the days of the state acquiring land to help ski areas begin or building access roads to connect them to state highways.

There are a number of types of second homes in the state. As discussed above, during the 1960s and 1970s there was a dramatic increase in the building of second homes and condominiums near ski areas. Indeed, such developments were the catalyst for Act 250. In the 1990s, they were concentrated at the major ski resorts, such as Killington, Stowe, Stratton, and Sugarbush. The other type of second home was most typically an existing house—often one in disrepair that needed significant renovations. A number of villages, such as Woodstock, Manchester, and Grafton, became second-home centers. Other such homes were in the country, many on rundown farms or in more established recreational settings along the state's lakes and ponds. Although the refurbishment of these existing homes meant no new construction, and hence did not alter the physical landscape, the arrival of wealthy outsiders often dramatically altered the cultural landscape due to increased property values and differing personal values. Overall, the number of second or seasonal homes increased significantly, from 22,000 in 1968 to over 47,000 thirty years later.

The interstates provided an additional boost to the growth of tourism in Vermont. As it took less time to get to Vermont from the surrounding areas, the state became a mecca for the long-weekend getaway. This kind of trip was popular virtually any time of year except mud season. A couple or family might come in the summer to do some biking, take in a concert, hike part of the Long Trail, and go antiquing. The fall trips focused on the foliage; the winter ones on downhill and cross-country skiing.

Beyond downhill skiing, other major landscape-related tourism draws to Vermont included cross-country skiing, snowmobiling, golfing, mountain biking, hiking, horseback riding, summer camps, hunting, and fishing. The latter five activities had been well-established aspects of recreation prior to 1950, but the first four were new. Cross-country-ski areas began to open in the 1970s, most affiliated with an existing downhill area or an inn. Overall, they had a very limited effect on the landscape, as they spawned virtually no additional development. The main effects of snowmobiles on the landscape, however, were disturbance of wildlife, air pollution, and the construction of new trails. Golfing and, more recently, mountain biking became popular as a result of efforts by Vermont entrepreneurs to attract new recreationists—and often in efforts by ski areas to create year-round resorts. Vermont had over fifty golf courses in 1995, many located in ski towns. Many of these same ski areas also catered to mountain bikers by opening their slopes to them in the summer. The landscape effects of both of these activities, though significant, tended to be localized—for example, fertilizer and pesticide use on golf courses and soil erosion from mountain-bike trails.

Work on the Long Trail and its corridor continued during this period, and its use grew as well. The Green Mountain Club's major focus was securing protection for the trail corridor by working with the state and the U.S. Forest Service to acquire land and by making agreements with private landowners—usually recreation easements—to achieve indefinite access to the trail. As use of the trail system accelerated, especially in the 1960s, the Club began to worry about overuse of the land. In response, naturalists were employed on the summits of Mount Mansfield, Camel's Hump, and Mount Abraham during the summer to help protect the fragile alpine vegetation on these peaks.

The themes of Vermont tourism launched in the decades preceding 1950 remained the foundation for tourism and much more in Vermont. The two central themes—mountains and their skiing and the pastoral landscape—played a major role in shaping the kind of tourism and the kind of development that came to Vermont, as well as in shaping government policy (for example, the banning of billboards in 1968). Furthermore, they helped to shape an image of Vermont that drew year-round residents and new economic ventures focused on quality of life. These themes and policies were also tremendously successful. In 1997, for instance, tourism was the number-two industry in Vermont (behind manufacturing), with an estimated twenty-seven million annual visitors. In the next two sections we will look at the components of this pastoral landscape—farms and forests—and at the evolution of conservation policy in Vermont, policy that was very much designed to protect this pastoral landscape.

The Components of the Pastoral Landscape

The trend of agricultural decline that had begun in the nineteenth century continued unabated into the present period. Indeed, according to many measures the decline accelerated. Compared to the previous sixty years, the decline in farm acres per year tripled from 1940 to 1980 (to over 50,000 acres annually), and the number of farms fell from 24,000 to approximately

6,000 in 1995, with fewer than 2,000 dairy farms left by 1997 (see table 6.2). Dairy was still the epicenter of Vermont farming: From the mid-1960s through the mid-1990s, dairy production constituted 75 percent or more of the state's agricultural revenues, a level of dairy prominence unequaled in the nation. Agriculture remained important to the state, accounting for about one fifth of Vermont's revenues in the 1980s. The Champlain Valley and north-central Vermont remained the center of dairy farming, with Franklin, Addison, and Orleans Counties leading in dairy production. Almost all of this milk was consumed out of state; less than one tenth was consumed in Vermont. Cheese enjoyed a real renaissance. Production had fallen to five million pounds in 1950, but by the mid-1990s cheese production exceeded 130 million pounds, the highest level ever.

Fully integrated into global agricultural markets, Vermont farmers relied on freshness and quality as their chief selling points. The former, as in the past, was based on the state's proximity to large urban markets. The latter was becoming more important as transportation and chemical advances, often financed by large-scale corporate agribusiness, reduced the freshness advantage. Vermont farm products, led by Ben and Jerry's ice cream, earned a reputation as wholesome, high-quality products for which people were willing to pay premiums. By receiving these premium prices, Vermont farmers competed beyond the fluid-milk market.

Due in part to this dairy focus, Vermonters were importing over 80 percent of their food products. In the middle 1980s, Vermont was importing 98 percent of its beef, 95 percent of its onions, 91 percent of its carrots, 64 percent of its potatoes, and 43 percent of its apples. This was far removed from the self-sufficient farmers of the early nineteenth century.

This general decline in agriculture became such a concern that by the 1970s, the state government and many groups in the state sought to initiate programs to conserve farmland. Although the average size of a farm and of

TABLE 6.2.
Farmland and Forestland in Vermont, 1960–1995

Year	Farmland (acres, % of state)	Forestland (estimated acres, % of state)
1960	2,945,343 (50) [1959]	4,230,000 (71) [1963]
1970	2,220,000 (37)	4,391,000 (74)
1980	1,740,000 (29)	4,512,000 (76) [1977]
1990	1,510,000 (25)	4,509,000 (76) [1987]
1995	1,370,000 (23)	4,538,000 (76) [1992]

Note: There is an overlap between farmland and forestland due to farm acreage in forest (usually cited as woodland).

a dairy herd increased over this time, the increases could not overcome the sheer loss of farms, and the percentage of land in farms was down from over 70 percent of the state in 1920 to less than one quarter by the middle 1990s. But the loss of farms, farm acres, farm laborers, and cows was more than offset—in a strictly economic sense—by increases in productivity, accomplished by technological change (e.g., pesticides, hormones, antibiotics), better breeding via genetics, and better feeding. The amount of milk produced in the state increased from 1.3 billion pounds in 1951 to 2.5 billion pounds in the early 1990s.

Outside of dairy, Vermont played a minor role on the national agricultural scene. In the mid-1990s it was forty-fourth among the states in gross agricultural receipts (trailing Connecticut and Maine in New England). Vermont maple-syrup production declined by roughly 80 percent from 1935 to 1960, reaching its low point. It has rebounded since, and the state still led the nation in production in 1995 (though it is greatly outproduced by neighboring Quebec). Apples were an important minor crop in the state through the period but were in decline by the middle 1990s.

As this brief discussion indicates, the importance of agriculture in the state changed dramatically since the middle of the nineteenth century. By 1995, agriculture no longer defined Vermont the way it once did. It was still economically important, but in many ways it had become just as important symbolically, as part of the state's heritage and as a fundamental piece of its pastoral landscape. Politicians still worked to develop programs to help dairy farmers (such as the Northeast Interstate Dairy Compact, enacted in 1996), and farmers themselves carved out entrepreneurial niches for themselves through specialty products or community-supported agriculture, but unless there is a dramatic change in the world, the likelihood of agriculture again dominating the Vermont landscape is slim.

The proportion of the Vermont landscape covered by forest continued to increase after 1950, as it did throughout New England. By the early 1980s, approximately three quarters of Vermont was forested, the same percentage as the rest of New England (see table 6.2). The forests that returned, though, were far different from the forests of the 1600s. In place of the wild forest of that time, forester Lloyd Irland divided the present New England forest into five categories: industrial, rural, recreational, suburban, and wild. The industrial forest, most common throughout Maine, spilled over into Vermont's Northeastern Highlands. This forest was primarily owned in large blocks by corporations or wealthy families, often based outside the region. The focal point of this forest was to produce fiber for profit, and these woods were constantly being transformed by the latest technological innovations, such as whole-tree chippers and harvesters. In 1997, Vermont passed laws banning the aerial and ground application of herbicides and regulating cuts over forty acres in size to try to limit the spread of these industrial methods in the state's forests.

Most of Vermont was classified as rural forest: a landscape of farms, forests, and rural residents. These rural forests had a distinct farming ancestry; they are typically old farm woodlots that expanded, as well as abandoned fields that were once cleared. These forests were just as likely to be unmanaged as managed. If managed, goals other than timber production—such as aesthetics or wildlife habitat—were frequently more important to landowners. Significant acreage was under the state's current use program, which required a forest-management plan. Unfortunately, only those who wished to actively manage the trees in the forest were eligible for the program; those who wished to set aside their land as wild forest must pay full taxes on the land. Ownership patterns within this rural forest also changed greatly. In 1950, Vermont farmers owned about half the commercial forestland in the state; by the mid-1980s, that figure had dropped to one fifth.

Industrial and rural forest categories constituted over three quarters of New England's forests. In Vermont, though, the recreational forest played an important role, especially in the Champlain Valley and Green Mountains. In the words of Irland, "The recreational forest includes those areas in which the forest land is principally a green backdrop for resorts and second homes. The ownership pattern has become so altered by subdivision, nonresident ownership, and scattered development as to make future use of the land for commercial timber, recreation, or other purposes difficult or impossible." The new roads and houses were of even greater concern for ecological reasons since they fragmented and diminished natural habitats and communities, points elaborated on in chapter 7.

Although Irland classified none of Vermont as suburban forest, one might argue that in the years since his book was written, much of Chittenden County and some of the adjoining counties entered this classification due to the growth and sprawl centered around Burlington. As for wild forests, small pockets were scattered throughout the state and the region, almost exclusively on public lands, and they will be discussed below.

The forest-products industry remained important in Vermont, even though the state had little industrial forest. There were important changes, however. Here, too, technology had significant effects—the chain saw and motorized skidders being the most important, though whole-tree chip-harvesting operations have grown quite significant recently. Due to external market forces, the demand for Vermont forest products declined from the 1910s through the 1970s as the state and the rest of New England could not effectively compete with producers from the Northwest and South. Furthermore, after 150 years of heavy cutting, Vermont forests had low-quality trees for lumber. These market forces were yet another crucial factor in the return of the Vermont forest; that is, spared by the lack of demand, Vermont's forests could grow and expand. The forest-products industry in the state began to rebound in the 1970s as quality trees became ready to harvest, and, in response to increased hardwood demand, whole-log exports to Quebec rose, as did increased fuelwood cutting because of the energy crisis. For the first time since the late nineteenth century, there was a growing demand on Vermont's forests for fuel. Firewood consumption increased by five times from 1975 to 1981. After peaking in the 1980s at over 400,000 cords per year, residential fuelwood harvests dropped to an average of 330,000 cords per year. In 1997, approximately 10 percent of the wood harvest in Vermont went to fuel.

That Vermont was fully integrated into the global forest economy was illustrated in 1987, when Korean firms imported one million of the twenty-nine million board feet of maple cut in the state (much of the demand was for making pianos). Overall, the annual cut averaged over 240 million board feet in the 1950s, but fell into the 170–180 million board-feet range in the 1960s and 1970s. Cutting rose again in the 1980s, averaging nearly 210 million board feet per year, then rose even further to nearly 270 million board feet per year in the 1990s. Pulpwood harvesting also increased significantly, from fewer than 40,000 cords in 1926 to an annual average of 389,000 cords in the 1990s. Whole-tree chip harvesting increased even more dramatically, from 34,000 cords per year in the 1970s to over 275,000 cords per year in the 1990s. Overall, timber harvesting in Vermont reached record levels in the 1990s, when all forest products—sawlogs, veneer logs, pulpwood, bolt wood, and whole-tree chips—were included. Indeed, the overall New England cut between 1980 and 1995 was at or near record levels. Through the 1980s and into the 1990s, the wood-products industry (including lumber, furniture, and paper) remained an important one in Vermont. It was the second-largest manufacturing industry in the state, though it was a minuscule part of the national industry, supplying roughly 1 percent of the national total of forest products.

The forest continued its return to Vermont through roughly 1980, when the amount of the state forested leveled out in the 75 to 80 percent range. Even though Vermont was approximately four fifths as forested as it was prior to the arrival of Europeans, the forest was biologically very different (see chapter 7). Furthermore, the social forces on the forest were also quite different, with the present forest shaped by residential fragmentation, tourism and recreation, global market forces, disease and pollution, and a myriad of conservation policies. As with almost all the Vermont landscape, major forces shaping the forest often arise beyond the state's borders. Disputes with Canada over such issues as whole-log exports and lumber-industry subsidies underscored the importance of political borders, even as the North

American Free Trade Agreement (NAFTA) ostensibly levels the economic playing field (and the environmental playing field with its Commission on Environmental Cooperation). NAFTA could lead to more raw-material exports from Vermont, since it seeks to remove all trade barriers between the United States, Canada, and Mexico. Such international trade policies will have significant—and likely unintended and unpredictable—consequences on Vermont's forests.

Protecting the Vermont Landscape: Public Lands and Conservation

Since 1950, the programs and agencies designed to foster the conservation of the landscape evolved and grew. The state and federal governments kept purchasing land for conservation reasons, while the years after 1950 saw the declining importance of municipalities in buying land for conservation reasons and the rise of private, nonprofit conservation groups—especially land trusts—in the acquisition of land and protective easements.

The town or municipal forest movement, which had such vigor in the first half of the twentieth century, lost momentum with the rise of the environmental movement in the 1960s and 1970s. In 1960, more than sixty-five towns had municipal forests covering over 33,000 acres. Throughout New England, states passed laws authorizing the establishment of local conservation commissions. These commissions—the first of which was authorized by Massachusetts in 1957, the last by Vermont in 1977—focused on broader environmental concerns at the local level, and in many towns such commissions took over the municipal forests. Indeed, soon after Vermont's enabling act was passed, the state's municipal-forest program was ended. Furthermore, as towns' needs changed, so did their desires to own forest parcels. Illustrative of this was municipal watershed forestland in Addison County. As communities developed more attractive water supplies (such as deep wells), they sought to sell watershed lands that were often located in different towns. Vergennes sold its watershed forest in Bristol to a local conservation group—the Watershed Center—in 1997; Middlebury sold its watershed lands in Bristol to a timber company in 1998; and Bristol would like to dispose of its no longer necessary watershed lands in Lincoln. These municipal lands covered more than 40,000 acres statewide in 1995 (see table 6.3).

The state forest system grew steadily. In 1959, there were twenty-eight state forests covering more than 80,000 acres. By the mid-1990s, there were thirty-seven state forests including 156,000 acres (see map 6.3). The state park system also grew at a healthy rate. There were twenty-eight such parks in 1959 covering more than 7,000 acres. Today, there are fifty-three state

TABLE 6.3.
Conservation Lands in Vermont, 1997

Landowner	Acreage	% of State
Federal agencies		
Forest Service	366,413	6.17
Fish and Wildlife Service	6,345	0.11
National Park Service	8,529	0.14
Army Corps of Engineers	5,806	0.10
Federal total (fee)	387,093	6.52
Federal nonfee	5,456	0.09
Federal total (overall)	392,549	6.61
State Categories		
State forests	156,039	2.63
State parks	39,859	0.67
Wildlife management areas	93,141	1.57
Other	5,770	0.10
State total (fee)	294,809	4.97
State nonfee	41,994	0.71
State total (overall)	336,803	5.67
Municipalities	40,571	0.68
Public fee	722,473	12.17
Public nonfee	47,450	0.80
TOTAL PUBLIC	769,923	12.97
Vermont Land Trust		
Fee	31,938	0.54
Nonfee	112,750	1.90
The Nature Conservancy		
Fee	10,550	0.18
Nonfee	450	0.01
Green Mountain Club		
Fee	3,736	0.06
Nonfee	810	0.01
Other private nonprofit groups		
Fee	8,174	0.14
Nonfee	2,018	0.03
Private fee	54,398	0.92
Private nonfee	116,028	1.95
TOTAL PRIVATE	170,426	2.87
Overall fee	776,871	13.09
Overall nonfee	163,478	2.75
TOTAL	940,349	15.84

Note: Nonfee acreage in almost all cases refers to conservation easements held on these properties. Which activities are allowed on these lands are subject to individual agreements, though in almost all cases no building or development of any kind is allowed. Often, agriculture and forestry operations are allowed. The amount of land protected in some categories is estimated. It is very likely that municipalities and private nonprofit groups have protected more acreage than is listed.

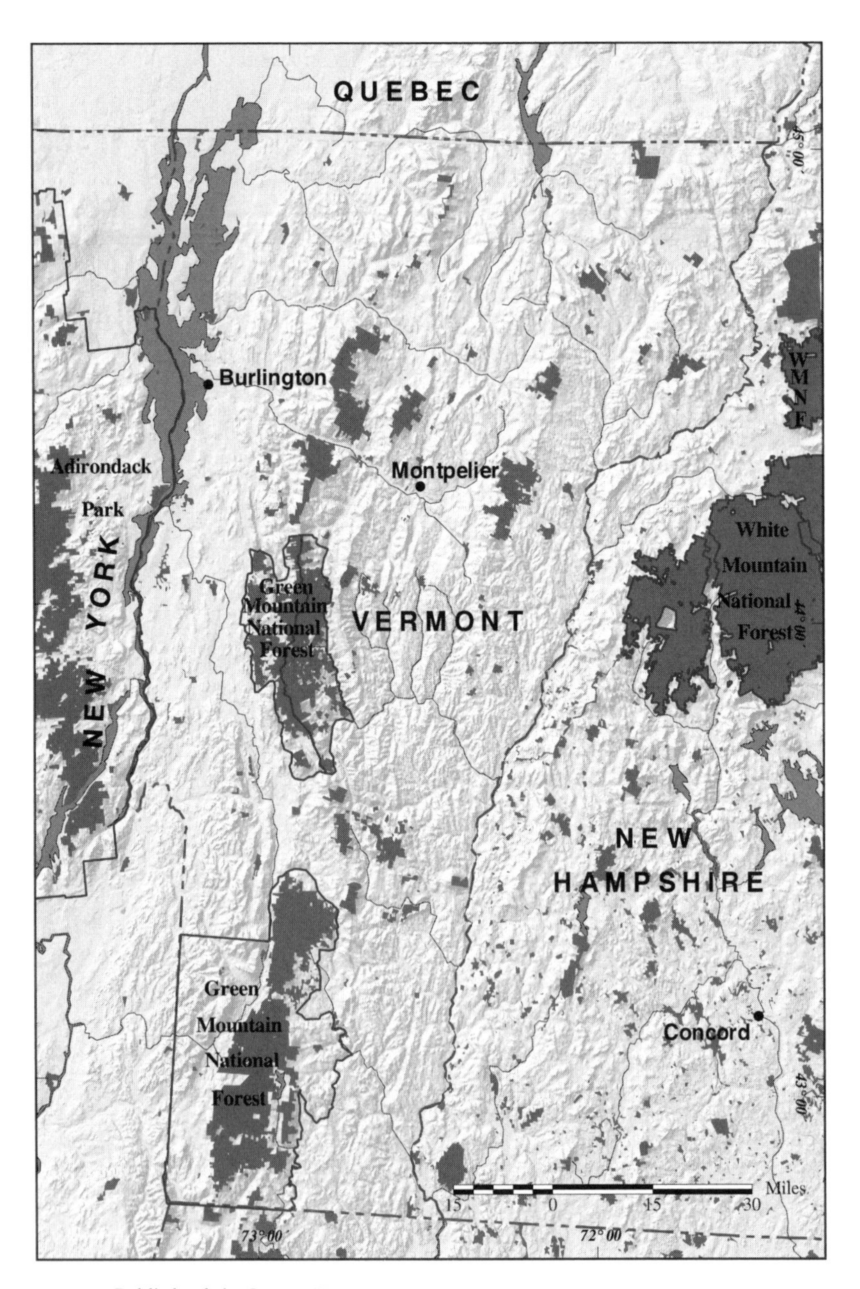

MAP 6.3. Public lands in Greater Vermont, 1995.

parks totaling nearly 40,000 acres. By the mid-1990s, the state had acquired over 290,000 acres of land in wildlife management areas, pond sites, stream banks, and boating-access areas, as well as in state forests and parks. The land was scattered throughout the state, mostly in small parcels. The largest areas were the Mount Mansfield State Forest (nearly 40,000 acres), the Groton State Forest (over 25,000 acres), and Camel's Hump State Park (roughly 18,000 acres).

State land acquisition was greatly aided from 1965 to 1987 by the Land and Water Conservation Fund (LWCF), a federal program initiated in 1964 to help all levels of government purchase land for conservation and recreation purposes. As the funding for this program declined drastically in the 1980s, Vermont started its own program to help finance such purchases. In 1987, the Vermont Housing and Conservation Trust Fund was established to help people purchase affordable housing, to protect farmland from development, and to aid nonprofit private groups in the acquisition of conservation and recreation land. Among the chief private groups involved in land acquisition through this program were the Vermont Land Trust, the Vermont Nature Conservancy, and the Green Mountain Club. In addition to this fund, the state also appropriated money directly for purchase of conservation and recreation lands.

This new public-private partnership in land conservation was the greatest innovation in land conservation in the modern era. Although the federal and state governments were still purchasing land to protect it (whereas local governments today seem to be more inclined to dispose of land rather than purchase it), an array of groups was also at work purchasing land that they retained, sold, or donated to the state or federal government, as well as purchasing development rights or conservation easements on land.

Three main types of groups were involved in this land protection. The first was the major land trusts, including the Vermont Land Trust, the Lake Champlain Land Trust, and the Upper Valley Land Trust (working in the upper Connecticut River Valley in both Vermont and New Hampshire). These trusts—beginning in the late 1970s—worked to protect both farmland and forestland from development, usually by purchasing development rights. These major land trusts were mainly concerned with protecting open space and maintaining a managed landscape. The Vermont Land Trust, the largest in the state, partially protected over 140,000 acres from 1977 to 1997. A second type of group was the local land trusts, such as the Middlebury Land Trust. These groups tended to have missions similar to those of the larger land trusts but worked in more specific regions. The third type of group had a more specialized mission in its land protection. The two most important such groups in Vermont were the Nature Conservancy and the Green Mountain Club. The Nature Conservancy purchased land throughout the state "to preserve the plants, animals and natural communities that represent the diversity of life on Earth." The organization helped preserve over 100,000 acres in Vermont and ran its own small reserve system. As discussed in the previous section, the Green Mountain Club focused on protecting the Long Trail corridor. In late 1997, the Vermont Land Trust and the Nature Conservancy combined to purchase nearly 27,000 acres of land for use in sustainable-forestry operations.

The amount of land in the Green Mountain National Forest increased as well. In 1959, the proclamation boundaries stood at 580,000 acres, with about 220,000 acres (38 percent) under federal ownership. The LWCF greatly helped accelerate land purchases for the national forest as well. By 1981, the proclamation boundary area had increased to 629,000 acres, with 290,000 acres (46 percent) owned by the federal government. Sixteen years later, the boundary had reached 815,000 acres (including a significant expansion of the forest to include the Taconic Mountains in southwestern Vermont in 1990) and federal ownership had grown to 366,000 acres (45 percent).

The Wilderness Act, a national law passed in 1964, substantially affected how roadless national-forest lands were managed across the country. The law defined wilderness as areas "where the earth and its community of life are untrammeled by man, where man himself is a visitor who does not remain." No areas in the Green Mountain National Forest were originally designated as wilderness, but in 1975 the Eastern Wilderness Act established the Bristol Cliffs and Lye Brook Wilderness Areas. In the Forest Service's second Roadless Area Review and Evaluation in the late 1970s, six additional de facto wilderness areas totaling over 55,000 acres were studied in the Green Mountain National Forest. The Forest Service recommended none of these areas for wilderness. This proved unacceptable to Vermont wilderness advocates, and they convinced the Vermont congressional delegation to back a plan to designate most of these areas as wilderness. The 1983 Vermont Wilderness Act did just that, creating four new areas (Big Branch, Bread Loaf, George Aiken, and Peru Peak) and expanding one (Lye Brook), bringing total designated wilderness in the state to just under 60,000 acres. It also created the 36,000-acre White Rocks National Recreation Area. The largest of the wilderness areas, Bread Loaf, was over 21,000 acres.

Management of the national forest grew increasingly controversial in the last decade. Debates over where and how much timber—if any—should be cut, what kinds of recreation would be permissible, and how much land should be protected as wilderness areas or in other ways were the central questions that made management of the Green Mountain National Forest increasingly complex and prone to conflict.

Another arena in which the federal Forest Service played a role in Vermont recently is the Northern Forest, a region demarcated in 1988 in response to a series of corporate takeovers of forest-products companies and subsequent land sales in Maine and the northern portions of New Hampshire, Vermont, and New York. The U.S. Forest Service's Northern Forest Lands Study (NFLS) and the Governors' Task Force on Northern Forest Lands studied land ownership, conservation strategies, forest resources, economics, and the human communities in the twenty-six-million-acre region, including two million Vermont acres in five northern counties. After these entities completed their studies in 1990, the Northern Forest Lands Council (NFLC) was created to complete further studies over another four years. Its studies were expanded beyond forestland conversion to also include biological diversity, forest health, and forest practices. The recommendations from the NFLC are still being considered, but overall these studies initiated a wide-ranging discussion about the forests of northern New England and New York. Furthermore, it was clear that the catalyst for these studies—major forestland sales—were still of major concern, as was illustrated in the fall of 1997 when Champion International announced that it was selling its 133,000 acres of forestland in Essex County in northeastern Vermont (as well as over 185,000 acres in the Adirondacks and northern New Hampshire).

A new federal program that came out of the NFLS was Forest Legacy. Administered by the U.S. Forest Service, the program provided funds to purchase conservation easements on forestland at risk from development. The lands remained in private ownership, but certain uses were no longer allowed. The landowner received compensation for surrendering certain rights—usually development rights. Depending on the particular parcel, the lands may be still used for timber harvesting, recreation, or conservation. The first use of the program was to protect 1,600 acres in Granby. More recently, over 30,000 acres of land in the Northeastern Highlands were protected via the Forest Legacy program.

The federal and state governments were also involved in another set of conservation-easement acquisitions, these dealing with hydroelectric-dam relicensing. Many such dams required new permits from the Federal Energy Regulatory Commission in the 1990s. As part of the negotiations to grant the permits, two major land-protection deals were completed. In February 1997, the state received permanent conservation easements to 16,000 acres along the Harriman and Somerset Reservoirs and the Deerfield River in southern Vermont. In September 1997, the state concluded a similar deal to protect four thousand acres along the upper Connecticut River.

Two other federal conservation agencies—the Fish and Wildlife Service and the National Park Service—were landowners in the state. The Missisquoi National Wildlife Refuge, established in 1943, covered approximately

6,300 acres located where the Missisquoi River empties into Lake Champlain in northwestern Vermont. The refuge was created and managed primarily for migrating waterfowl. The recently established Conte National Wildlife Refuge, which will have landholdings throughout the entire Connecticut River Valley, planned to acquire—and to help state and local organizations acquire—over 25,000 scattered acres in Vermont. Another recently created federal conservation unit is the Marsh-Billings National Historical Park in Woodstock, authorized in 1992. The park, less than one thousand acres, honors the birthplace of George Perkins Marsh and the conservation legacy of both Marsh and Frederick Billings. Total federal ownership in 1995 was 6.6 percent of the state's land, compared to the 3.4 percent the federal government owned overall in New England.

Along with the continuing importance of public-lands acquisition in Vermont conservation, state administrative capacity to deal with conservation issues developed further. In 1955, the State Forest Service was upgraded to the Department of Forest and Parks. Six years later, the Fish and Game Service was similarly upgraded, and a Department of Water Resources was established. A major reorganization took place in 1970, when existing natural-resources agencies and new environmental ones were placed in the Agency of Environmental Conservation. The final major change came in 1987, when this agency was renamed the Agency of Natural Resources. It consisted of three main departments: Forests, Parks, and Recreation; Fish and Wildlife (the change from game to wildlife came in 1983); and Environmental Conservation (which administered pollution-control policies). Various advisory boards were created as well, on topics such as forests, energy policy, Lake Champlain, and nuclear issues.

The state and federal governments also played major roles in wildlife policy. Until the 1970s, their roles had concerned managing animals for hunting and fishing. Although there was some decline in the sale of licenses for hunting and fishing over this time, those activities remained extremely popular in Vermont. Indeed, in 1996 over 225,000 hunting and fishing licenses were sold. Perhaps the biggest change in wildlife management during this period took place in 1966, when the state legislature, which had established the seasons until that time, granted control of the deer hunt to the Fish and Game Department. During a twenty-year period of controversy over the size and health of the deer herd, the legislature took back control of the hunt from the department in 1971 and then returned it again in the late 1970s. This department—with the Fish and Wildlife Board—controlled all hunting and fishing in the state, including the reintroduced moose hunt in portions of northern Vermont. The department also ran a series of fish hatcheries, totaling five in 1997 (along with two federal hatcheries), to stock the state's waters.

In addition to hunting and fishing management, endangered species became a major government concern, beginning with the 1973 federal Endangered Species Act. As of 1995, five animal species native to Vermont were listed on the federal endangered species list (Indiana bat, mountain lion, peregrine falcon, bald eagle, and dwarf wedge mussel), as well as one threatened animal (Puritan tiger beetle). Two plants were listed as endangered (Jesup's milk vetch and barbed-bristle bulrush) and one as threatened (small whorled pogonia). The state also had an endangered-species law, enacted in 1981. Under this law, twenty animals and sixty-one plants were listed as endangered, and fourteen animals and ninety-two plants were listed as threatened. Thus far, these laws have not led to the kinds of controversies that have engulfed the Pacific Northwest over the spotted owl and salmon. As discussed in subsequent chapters, the Fish and Wildlife Department was involved in reintroducing a number of species to the state for hunting purposes (e.g., wild turkey) and endangered-species purposes (e.g., peregrine falcon).

As the forests of Vermont returned, as the state's population greatly increased, as Vermont became more and more integrated into a global economy and society, Vermonters acted—both privately and publicly—to protect the landscape. These efforts covered a wide spectrum, from federal wilderness to conservation easements to protect working farms; from endangered species laws to active game and fish management. But in some sense, these actions (except those dealing with agricultural lands) have been trying to reassemble a natural-landscape mosaic out of one that had been changed completely by the mid-1800s. Though nearly 16 percent of Vermont's land became protected to some degree, much of this land is still greatly fragmented. One conservation lesson society is still learning—in Vermont and elsewhere—is that it is a lot harder to put something back together than it is to take it apart.

Industry, Energy, and Pollution

Vermont remained an industrial backwater through the 1960s. The remaining textile mills gradually went out of business following World War II, unable to compete with mills in other parts of the country. Even those mainstays of Vermont industry that remained in business—such as the Vermont Marble Company and Fairbanks-Morse Scales—became part of corporations based out of state. The decline in Vermont mining, which had begun in the early twentieth century, continued—with the exception of a flurry of activity during and immediately after World War II—and was even more pronounced than that of agriculture. In the mid-1970s, Vermont led the nation in talc production, was second in marble, slate, and asbestos production, and was third in granite production. But the nation's demand for these products was declining. Synthetics curtailed the demand for talc; asbestos demand plummeted as the mineral became associated with cancer; and Americans demanded less stone for construction than they once did. Other mineral products in the mid-1970s included kaolin, dolomite, limestone, soapstone, aggregates, and sand and gravel. These trends in mineral demand continued into the 1990s. Though Vermont remained second in asbestos production, third in talc production, and fifth in dimension-stone production (granite, marble, and slate), the production of all of these minerals has declined substantially over the last twenty years.

The new industries that did come to Vermont during this time were the branch operations of multinational corporations such as General Electric, St. Regis Paper (the largest private landowner in the state circa 1980), and IBM. The latter, locating in Essex in 1957, soon became the driving force in the economic and population growth of Chittenden County. Within eight years, IBM became the largest employer in the state, and was still so in 1997. By the 1970s, Vermont manufacturing, service, and trade were all fundamentally tied to economic trends outside of Vermont. Furthermore, the new high-technology firms did not need large amounts of raw materials, nor did they ship heavy products, so Vermont's distance from most population centers was not as problematic as it had been. Several such high-technology firms, including IBM, were also interested in locating in perceived high-amenity areas in order to attract better workers.

Energy demand increased dramatically since 1950, with the rise of industry in the state and, most important, the sharp rise in population. Vermonters, like Americans throughout the country, consumed high levels of energy for transportation, heat, and electricity. Through the middle of the nineteenth century, Vermont had been self-sufficient in energy, relying on wood and water for heat, to cook, and for power. Fossil fuels then entered the scene, with coal replacing wood as a source of energy. Oil and natural gas followed by the end of the century. Since Vermont had no fossil fuels, the state rapidly became energy dependent, yet again at the whim of forces beyond its borders. In response to the energy crisis of the 1970s, many Vermonters returned to the state's original power source: wood. Around 1980, wood supplied roughly 15 percent of Vermont's energy needs, with over 40 percent of houses burning wood as the main source of heat. Both of these figures had declined by roughly half by 1993: wood supplied 7 percent of the state's energy needs, and it served as the primary heat source in 21 percent of households (and the supplemental source in 17 percent). In 1993, petroleum products for transportation and home heating made up 71 percent of the state's total delivered energy use. Natural gas contributed another 6 percent.

The other 16 percent of delivered energy use in 1993 was in the form of electricity. Due to the complex web of electric generators and transmission systems—with much of the electricity produced in Vermont consumed elsewhere, and much of the electricity consumed in Vermont produced elsewhere—one needs to look at how the electricity in Vermont was produced and consumed. In terms of production in the state, over 80 percent of the electricity generated in the state in 1994 came from Vermont's one nuclear-power plant, Vermont Yankee, which began operating in 1973. Since Vermont utilities owned only 55 percent of the plant—located in the southeastern corner of the state—much of the energy generated there left the state. Overall, a third of the electricity consumed in Vermont is nuclear in origin.

Vermont still relied on water for power, at least to some degree. The first hydroelectric plant in the state opened in 1886, and the Comerford Dam on the Connecticut River, opened in 1930, was the fourth-largest dam in the country when it was built. By 1994, 17 percent of the electricity generated and 41 percent of the electricity consumed in the state came from hydro sources. Perhaps fitting for a period when political borders have become less meaningful, most of the hydro power generated in Vermont was not consumed in the state. New England Power had six dams on the Connecticut River (Vernon, Bellows Falls, Wilder, McIndoe Falls, Comerford, and Moore), but all of the electricity generated from them went to customers elsewhere. The same was true for the utility's two major generating stations on the Deerfield River in southern Vermont (Searsburg and Harriman), which formed Harriman and Somerset Reservoirs, two of the largest bodies of water in the state. The roughly seventy plants scattered on other rivers throughout Vermont in the late 1970s were relatively small operations. By the 1990s, many environmentalists were focusing on the negative consequences of even these small hydro operations. The major battle was over a breached dam on the Clyde River, flowing into Lake Memphremagog: the utility wanted to repair the dam; environmentalists and fishers wanted it removed. The latter won. Further controversies of this nature are likely in the future. (The ecological implications of such actions are discussed in chapter 9.)

A last note on hydroelectric power: It led to the final dispute over Vermont's borders. In 1915, Vermont sued New Hampshire, claiming the boundary between the states was the middle of the Connecticut River, rather than the west bank as claimed by New Hampshire. Vermont wanted to be able to tax the new dams that were to be built on the river. In 1933, the U.S. Supreme Court ruled that the border was the low-water mark on the western side of the river.

Just as electricity generated in Vermont served customers elsewhere, Vermonters received electricity generated by hydroelectric dams outside Vermont. The first major source was the New York State Power Authority; then, in the late 1970s, Vermont started to get power from Hydro Quebec (HQ). Major long-term contracts were signed by the state and its utilities with HQ in 1984 and again in 1989. In the middle 1990s, two thirds of the state's electricity generated by hydro came from HQ. Here activities and decisions by Vermonters—and others—had significant effects on people and landscapes beyond the state's borders. Future proposed contracts with HQ touched off significant protest within the state for these reasons.

Renewables—mainly wood—played a smaller role in generating electricity in Vermont. The Burlington Electric Department completed the largest utility-owned wood-burning electric plant in the country in 1984. Since then, another plant opened in Ryegate. Given the relative lack of direct sunlight in Vermont (Burlington was the sixth cloudiest major weather station in the lower forty-eight states), the only alternative energy source that was much explored in the state was wind power. The first commercial wind-generating station in the country functioned for nearly a month in Rutland County in 1945 before it was knocked down by gusty winds and not replaced. More recently, in 1997 Green Mountain Power began operating a series of eleven wind turbines on a ridge in Searsburg, near the Massachusetts border.

Throughout the nation, the 1960s and early 1970s was a period of focus on environmental pollution. National laws to regulate pollution were passed, with the states playing a major role in implementing them. The Environmental Protection Agency (EPA) was created in 1970 to oversee this process at the national level; the Agency of Environmental Conservation was created the same year to implement these laws in Vermont. In general, the state received high rankings for the quality of its natural environment and its environmental-policy capacity. In 1991, for example, it ranked third in the Green Index of the states (compiled by the Institute for Southern Studies in North Carolina).

Water quality in Vermont was regulated by the state starting in 1949. Two years earlier, the legislature passed a law enabling municipalities to organize sewage-disposal districts. Further state action came in 1970 with the Water Pollution Control Act, which required that any entity discharging waste into a stream or body of water must receive a permit from the state Department of Water Resources. This permit system, modified by the 1972 Federal Water Pollution Control Act and its amendments, is still the basis for controlling point-source water pollution, that is, water pollution that can be identified as coming from a particular source, such as a pipe. With municipalities and industries constructing facilities to treat such wastes, the level of point-source water pollution in Vermont declined significantly since 1970. Thanks to nearly two hundred municipal and industrial wastewater-treatment facilities, water quality improved in nearly sixty rivers and four lakes. Indeed, only 10 percent of water pollution came from point sources in the mid-1990s. The bigger challenge in Vermont was nonpoint-source pollution, such as runoff from agriculture, forestry operations, septic systems, parking lots, lawns, and golf courses. The dispersed nature of such nonpoint sources makes them much harder to control. For instance, controlling phosphorus entering Lake Champlain was identified by the Lake Champlain Basin Program as one of the three major priorities for health of the lake; a 1994 study of the lake found that 47 percent of phosphorus entering it came from nonpoint sources (with 29 percent from point sources and 23 percent from natural sources). Another major water-pollution problem in Vermont results from toxic pollutants, for example, PCBs and mercury. Such pollutants were locally problematic, usually found in lake and river sediments. Sources included past industrial pollution and airborne deposition. The latter problem, like acid rain discussed below, was mainly because of pollution from beyond Vermont's borders.

Major efforts to control air pollution in Vermont and the rest of the nation can be traced to the 1970 Clean Air Act, which still serves as the framework for air pollution–control policy in the United States. Due to its relatively low and dispersed population, as well as the relative lack of industry, Vermont has not had significant problems with the six main ambient air pollutants (carbon monoxide, lead, nitrogen dioxide, particulate matter, ozone, and sulfur dioxide). In 1994, for instance, the levels of these six pollutants in Vermont air were below 50 percent of federal standards, and trends were improving (excepting ozone).

There were, however, a number of significant air-pollution problems in Vermont: acid deposition (referred to more commonly as acid rain), ozone, air toxics, and global-climate change. Acid rain is formed when sulfate and, to a lesser degree, nitrate pollutants create acids in the atmosphere, which then fall to the surface in precipitation. The major sources of these pollutants were coal-burning utilities and automobiles, primarily upwind from Vermont in the Midwest. As is further discussed in chapter 7, this precipitation acidified a number of water bodies in Vermont and the surrounding region and may have contributed to red spruce decline at higher elevations. The 1990 Clean Air Act amendments at the federal level included provisions to reduce the precursors of acid rain by half, but it is not yet clear if these changes will fully address the problem throughout the Greater Laurentian Region. Indeed, in the mid-1990s, total precipitation acidity levels had declined little compared to the 1980s. Ground-level ozone, or smog, was another air-pollution problem that was generated primarily beyond Vermont's borders. Again, electric utilities and automobiles were the chief culprits. Emissions from these sources, under the right atmospheric conditions, combined and were blown into Vermont by the prevailing wind currents. This smog caused significant human health problems, impaired visibility, and damaged certain plant species. As was the case with acid rain, the EPA and federal air-pollution policy worked to reduce this problem throughout the northeastern United States. Air toxics are a host of air pollutants that were less widespread throughout the nation and Vermont, yet they can have high local concentrations and significant health and ecological effects. Although Vermont had comparatively low levels of toxic releases (into air, soil, or water), and such releases declined by over 60 percent between 1988 and 1994, over 99 percent of the releases were into the air. In some places in Vermont, especially urban areas, toxic standards were exceeded for such pollutants as benzene and carbon tetrachloride. Furthermore, many of the toxic air pollutants in Vermont came from outside of the state. Both state and national air-pollution laws began focusing more attention on controlling such pollutants in the 1990s. Finally, global-climate change could have tremendous effects on the Vermont landscape. Based on some projections, the forest types in the state could not survive at higher temperatures, which could mean an entirely different natural landscape in Vermont. The scope and scale of this problem are well beyond the control of Vermonters; if such projections are accurate, the Vermont landscape will be completely at the mercy of the earth's systems and the actions of the rest of the people in the world, although Vermonters, too, are adding to the problem.

Beginning in the 1960s, Vermont sought to more effectively manage and reduce its solid waste. Municipal landfills were regulated in the late 1960s, followed by the 1972 state bottle bill, which required deposits on soda and beer bottles. Fifteen years later, the Solid Waste Act was passed, establishing a statewide framework for solid-waste reduction. In 1989, the state adopted a target of diverting 40 percent of the municipal waste stream from landfills via reduction, reuse, and recycling by 2000. At the time, 19 percent was being diverted; five years later, 33 percent was.

Given its relatively small population and lack of heavy industry, hazardous-waste sites have been less of a problem in Vermont than in many other states. Nonetheless, in 1997 eight sites in the state were on the EPA Superfund National Priorities List. Furthermore, in the same year Vermont had more hazardous-waste sites than dairy farms. Of the more than two thousand sites, three quarters involved leaks of underground tanks; at over one thousand of these sites treatment was completed or the leaks were negligible. The major hazardous chemicals in the state were pesticides, mostly used in agriculture. Over 500,000 pounds of pesticides were used on Vermont farms in 1996. Some health advocates focused on these chemicals and their role in Vermont's breast-cancer rate, eighth highest in the country. Vermont will also soon begin to export some of its most hazardous waste.

The state recently became part of a compact with Texas and Maine to ship its low-level radioactive waste to a site in Texas, despite the significant opposition of the people who live near the facility.

During this period, from 1950 through 1995, the amount of forested land in the state continued to increase, up to roughly three quarters of the state. Accordingly, the percentage of land in the state that was part of a farm continued its steep drop, to less than a quarter. This return of the forest in Vermont, though, did not reflect a return to simpler times or to a human-nature interaction like that of the Abenaki. Rather, this return was related to the complexities of markets, politics, and society extending far beyond Vermont's borders. As environmental historian Carolyn Merchant writes, "This dependence on outside markets has moved some types of environmental degradation beyond New England's boundaries, allowing portions of its own environment to recover." In Vermont, this was seen most clearly in the return of the forests. Even as Vermont's forests returned, they suffered from the fallout of industrial capitalism (chiefly in the form of acid rain). Meanwhile, the state's population increased and relied on places beyond its borders for more and more of its food and energy (a point we return to in chapter 10).

In chapters 7, 8, and 9 we turn our attention to the third main force shaping the Vermont landscape: ecology. The interaction of climate, plant species, animal species, and ecological processes combine to create distinct natural-community types: forest, terrestrial open, and aquatic. The preceding discussions have, for simplicity, treated the biological characteristics of the Vermont landscape—such as forestland, streams, and fields—as if they were uniform in character. This is not the case, however, and the diversity of the natural-community types is part of the story of Vermont's landscape and how it has changed over time; they have developed in concert with the cultural landscape of Vermont. The next three chapters flesh out the details of the natural landscape today based on the human forces described in chapters 3 through 6. Although humans play a major role in determining the distribution of ecological-community types, ecological forces have been the key to determining what the Vermont landscape has looked like since the ice sheet receded 12,000 years ago. Whether humans seek to manage the landscape for agriculture or forest products or simply let nature take its course, these ecological forces are constantly at work. Indeed, they make up the real working landscape.

PART III

Ecological Communities of Vermont

7

Forest Communities

IN THE PRECEDING chapters, we presented many of the details of the natural history of Vermont only briefly to preserve the story's narrative flow. These details are not unimportant, however, as they provide (1) the context to explain many of the possibilities of and limitations to Vermont's cultural history, (2) an appreciation of the impact of culture on natural landscapes, and (3) an understanding of the options available for future land-use decisions in Vermont. We noted earlier that efforts by society to conserve nature began in Vermont as early as the mid-1700s and have continued up to the present. We believe that the goals of conservation are best achieved by strategies that are based on an appreciation of the diversity of nature, the ecological processes that operate in nature, and the effects on nature of cultural activities.

Although the landscape of Vermont is made up of a wide variety of natural-community types, the dominant feature of this region is surely its forests. For much of the time since the woodlands and forests spread throughout the area more than 11,000 years ago, forests of one type or another have probably covered 95 percent or more of Vermont. Even today, forests are found over about 75 percent of the state, so an understanding of these communities is central to an understanding of the natural history of Vermont.

Natural-Community Types

A natural community is a collection of species that interact with one another and with their environment and that are relatively free from human disturbance. Since any square mile of Vermont probably contains thousands of species of plants, animals, fungi, and bacteria, it is impossible to define and characterize a community with complete precision. However, for the purpose of a general appreciation of the natural history of Vermont, its local communities can be described based on a few dominant species and key physical characteristics.

For example, if more than 30 percent of the space overhead is covered by treetops, the community is considered to be a forest. Within the category of forest are subcategories. A forest with an average of less than 60 percent canopy cover is a woodland; with greater than 60 percent canopy cover it is a closed forest. A closed forest with species adapted to moist conditions is a mesic forest; one with species adapted to dry conditions or frequent fire is a xeric forest. A mesic forest dominated by the hardwood-tree species American beech, yellow birch, and sugar maple is a northern hardwood forest. Subdivisions of communities, or community classifications, can be made at finer and finer scales based on more and more characteristics of the species and habitats that they comprise.

A number of different systems for classifying natural communities are in use today, and no single system is perfect. For the purpose of developing a general understanding of Vermont's forest communities, we use here a simple system that considers only broad patterns of species associations. Although a forest community contains species from all of the major forms of life—including plants, animals, bacteria, and fungi—its most obvious members are trees. The presence of a particular set of tree species often correlates with particular climate, soil, and natural-disturbance patterns (such as fire and flooding), as well as with other species in the community; trees are, therefore, good general indicators for the entire community.

Before looking at the forest communities of Vermont in more detail, it is useful to understand the interrelationship between the physical conditions at a site and the plants that live there. This relationship was seen in a general way in the transition from tundra to open woodland to closed forest following the end of the Wisconsin glaciation; as the ground thawed and the climate warmed, the most successful species changed from low shrubs and grasses to alder and spruce and then eventually to hardwoods (see chapter 2).

In general, three key characteristics influence the kinds of plants that live in a region. The first is climate, which is a combination of both temperature and water availability. Species differ in their tolerance for heat or cold and for wet or dry conditions. In fact, the distribution of the two great forest types—deciduous and coniferous—is based on this difference in tolerance. Deciduous species (collectively called hardwoods)—those whose leaves die in the autumn, such as maple, birch, beech, and oak—fare best where the temperature is relatively mild during most of the year and precipitation is either evenly distributed throughout the year or reaches a peak during the summer. Most deciduous trees require a growing season of at least 120 days with an average daily temperature of greater than fifty degrees.

With few exceptions, coniferous or evergreen species (called softwoods) have needles that do not die in the autumn, and therefore they have foliage throughout the year. Some conifers, such as spruce and fir, fare best in regions and local areas that are extremely cold in the winter. In contrast to deciduous trees, spruce and fir require only thirty days during the year with temperatures greater than fifty degrees.

Climate also controls the distribution of species within these groups. Hemlock, a conifer, prefers warmer temperatures than does spruce. American mountain ash prefers cooler temperatures than does sugar maple. In general, oak and hickory fare better in drier conditions than do maple, beech, and birch.

Community structure is also controlled to a large extent by disturbance. Many kinds of natural disturbances—including fire, wind, flooding, landslides, disease, and insect and fungal attack—influence which species will persist in an area. Forest-community types often reflect the long-term history of disturbances there. The steep topography of the mountainous regions of Vermont results in many rock slides, which are the sites of a number of communities adapted to these periodic disturbances. Because fires naturally tend to be small and infrequent in Vermont (see chapter 5), due largely to its wet summers, few species here are specifically adapted to grow where fires are frequent. The cones of red pine, however, open only when the air is hot and dry or there has been a forest fire. Also, red pine is a poor competitor against other tree species and does not grow well in shade. The distribution of this species in Vermont, therefore, is heavily influenced by fire, which helps to open up the canopy and allow red pine seeds to spread.

The third factor that influences community type is soil. Where soils are thin or low in nutrients, forests have trouble becoming established. It took almost two thousand years after the retreat of the glacier from southern Vermont for soils to develop to the point of being able to support forests. The structure and chemistry of soils also influence which species are dominant. Eastern white pine often dominates where soils are sandy. Where soils are on calcareous rock, such as marble and limestone, trees such as bitternut hickory, butternut, and basswood are common. A number of herbaceous plants, such as hepatica, blue cohosh, herb Robert, and wild ginger also tend to be found in mineral-rich soil.

The cause-and-effect relationship between tree species and soil condition works both ways. Clearly, soil type is strongly influenced by past geological history; the clays and marine sediments of the Champlain Valley are the result of the formation of Lake Vermont and the Champlain Sea 10,000 to 13,000 years ago, independent of what has grown there since. But plants

TABLE 7.1.
Forested Natural Communities of Vermont

Cold-climate forests
Spruce-fir forest
High-elevation hardwoods-spruce forest
Subalpine heath/krummholz forest
Northern/high-elevation talus woodland
Northern hardwood forests
Beech-birch-maple forest
Northern hardwoods talus woodland
Hemlock forest
Floodplain forest
Transition hardwood forests
Lake bluff cedar-pine forest
Oak-hickory–northern hardwood forest
Transition hardwoods talus woodland
Pine-oak-heath sandplain forest
Rocky ridge and cliff woodlands

Note: Modified from a classification system used by the Vermont Agency of Natural Resources.

exert their own influence on soil condition. Over time, coniferous trees make the soil more acidic than deciduous trees do. This, in turn, affects the animals and bacteria that live in the soil as well as the shrubs and herbs that grow under the canopy of the trees.

According to one classification system, at least thirteen different forest community types are found in Vermont (see table 7.1), in total containing over one hundred different species of trees and another 550 species of shrubs and herbs. Of course, these communities have changed over time even independent of recent modifications by humans—for example, sugar maple arrived here at least one thousand years later than did hemlock, and American chestnut has virtually disappeared in the last fifty years—but the essential characteristics of Vermont's forest communities, including their overall species composition, have remained fairly stable for at least the last 4,500 years (see chapter 2).

Cold-Climate Forests

At the highest elevations and wherever climate conditions lead to very cold winters, a group of forest communities well adapted to the cold are found. These communities in Vermont are located in the Northeastern Highlands, the upper elevations of the Taconic Mountains, and above 2,500 feet in the

FIGURE 7.1. A spruce-fir forest community. Photo by Stephen Trombulak.

Green Mountains. These forests are dominated by coniferous trees, particularly red spruce and balsam fir, that extend up to the ridgelines of the mountains except on the highest peaks and most exposed faces.

The most common of the cold-climate forest communities in Vermont is the spruce-fir forest (see fig. 7.1). Spruce-fir forests also contain a small number of individuals of other tree species, including white spruce, black spruce, paper birch, and yellow birch. It is a matter of some debate whether or not this community today represents true boreal forest. When considering the global distribution of boreal forests, most authors consider this forest to extend only thorough Canada, Alaska, northern Europe, and Siberia, with a southern boundary at roughly fifty to sixty degrees north latitude. On the other hand, based solely on vegetation type, a boreal forest is usually considered to begin where broadleaf deciduous trees become a relatively minor part of the forest, which is certainly true of the spruce-fir forests of the mountainous regions of Vermont. Perhaps the cold-climate forests of Vermont and other locations in the Appalachian and Adirondack Mountains are more appropriately considered as sub-boreal forests: large regions of northern conifers that, in general, persist south of their primary distributions in high-elevation climatic islands, surrounded on all sides by low-elevation deciduous hardwoods.

A number of variants of this community type are recognized, based on

FIGURE 7.2. A krummholz forest on Mount Abraham in the Green Mountains. Photo by Stephen Trombulak.

the relative abundance of different hardwood species. One community, the high-elevation hardwoods-spruce forest, is an intergradation between the spruce-fir forests and the predominantly hardwood forests of lower elevations. High-elevation hardwoods-spruce forests, common between 2,000 and 2,500 feet throughout the state, have relatively low species diversity compared to forest types more dominated by hardwoods and primarily contain birches, with spruce and fir in lower numbers.

On the highest peaks in the Green Mountains, where exposure to winter winds is extreme, a special forest community made of dense thickets of stunted and twisted spruce and fir develop. These thickets, dominated by trees that rarely grow taller than six feet and that are usually much smaller and shrubbier than upright trees, are called krummholz, a German word meaning "crooked wood" (see fig. 7.2). In krummholz, the winds are so cold and fierce that the portions of the trees that survive to grow from year to year are those that are protected under the snowpack; parts sticking up out of the snow are regularly broken off and killed by ice and wind. Krummholz forests often have scattered herbs and shrubs within them that are more characteristic of alpine communities—such as highland rush and bilberry; these are called subalpine heath/krummholz communities. Particularly good examples of this community are present on Mount Mansfield, Camel's Hump, and Killington Peak.

Another type of cold-climate forest in Vermont, northern/high-elevation talus woodlands, develops on and below rockfalls at the bases of major cliffs, such as the Great Cliff and Bristol Cliffs, both in the Green Mountains. The movement of cold air down these generally open slopes and the unstable ground of the loose rock create conditions that favor the development of unique talus woodlands. Which species dominate at these sites depends on the climate and type of rock that forms the talus, but they may include black spruce, paper birch, mountain maple, American mountain ash, and a host of herbs, mosses, and lichens. This community is rare in Vermont simply because talus slopes are uncommon and small.

Many herbs are found more commonly in cold-climate forests than elsewhere. These tend to be species adapted to prolonged cold, long-lasting snowpack and acid soils typical of coniferous forests. They include whorled aster, mountain sorrel, blue-bead lily, bunchberry, shining club moss, mountain wood fern, and twinflower.

Northern Hardwood Forests

The most common of all the forest types in Vermont is the northern hardwood forest (see fig. 7.3). This community is adapted to conditions intermediate between the extreme winter cold of more northern latitudes and

FIGURE 7.3. A northern-hardwood forest community. Photo by Stephen Trombulak.

the summer heat of more southern latitudes. The widespread distribution of northern hardwoods in this region is evidence that Vermont today lies within the broad ecological boundary between two of the major forest regions of the world: the northern boreal forests and the southern temperate broadleaf deciduous forests.

Northern hardwood forests are found in all six of the biophysical regions of the state except where the climate is too cold (particularly above 2,500 feet), too warm (especially in the Champlain Valley and the southern part of Vermont), or too frequently disturbed (especially by fire). They are dominated by sugar maple, American beech, and yellow birch but also contain a host of other tree species, particularly white ash, hemlock, basswood, white pine, black cherry, striped maple, butternut, red maple, northern red oak, mountain maple, and paper birch. This forest community has played the largest role in the cultural history of Vermont, providing fuel, building material, and food to humans since the time of the Archaic people. Several variants of this forest type are recognized, based on which species share dominance with the beech, birch, and maple. These variants are influenced largely by soil condition and climate but also over the past two hundred years by their management and disturbance history.

Where soil nutrients are abundant because of mineral-rich bedrock or till, or where slopes cause downhill movement and accumulation of nutrients, there is an abundance of cherry, basswood, white ash, and butternut. White pine, the tree so important to early colonists for the trade in ship masts, flourishes where the soil is sandy. Northern hardwood forests that grow where it tends to be warmer, such as in southern Vermont and in the Champlain Valley, have a greater number of northern red oak. Woodlands that develop on and below rockfalls at the bases of major cliffs where the climate favors deciduous trees have a greater abundance of red maple, eastern hop hornbeam, and mountain maple.

One variant of the northern hardwood forest found throughout the state is the hemlock forest (see fig. 7.4). Hemlock is not a hardwood but a conifer adapted to living at warmer temperatures than either spruce or fir. In many locations, particularly where conditions are cool and wet and the soil thin, hemlock grows along with northern hardwood species. Because hemlock canopies block much of the light from the forest floor throughout the year, the understory and herb layer are sparse and without many species. Pockets of hemlock forests are easy to see during the winter on the lower slopes of the Green Mountains, for example, because the green of the hemlock trees stands out against the brown and gray of the leafless hardwoods. These pockets of hemlock provide important protection during winter for many animals, especially white-tailed deer.

Another variant of the northern hardwood forest, the floodplain forest,

FIGURE 7.4. A hemlock forest community. Photo by Stephen Trombulak.

represents an intermediate between terrestrial and aquatic communities in terms of the amount of water that covers the ground. Floodplain forests develop along the shores of rivers and lakes in areas prone to flooding, even if only for a short period of time in occasional years. They are found widely along all of the major rivers in the region as well as along the shore of Lake Champlain and share many characteristics with both well-drained terrestrial forests and forested swamps. The dominant floodplain forest tree in Vermont is silver maple, but other species are found there as well, including eastern cottonwood, basswood, swamp white oak, black willow, butternut, white and green ash, sycamore, hackberry, and occasionally elm. Many different species of herbs can be found there, especially ostrich fern, sensitive fern, and wood nettle. Regular flooding there adds nutrients to the soil, making the sites of floodplain forests good for agriculture. Since the earliest times of human agricultural settlement nine thousand years ago, but especially during the last three hundred years, floodplain forests have been extensively cut down and converted to farmland.

The annual turning of hardwood-tree leaves in the autumn provides Vermont with one of its most characteristic natural sights: fall foliage. After having grown and retained their leaves all spring and summer for photosynthesis, during the fall hardwoods begin to prepare for winter dormancy. As the days get shorter, chlorophyll—the molecule in the leaves that aids in photosynthesis and gives the leaves their green color—begins to break down. Once the chlorophyll is gone, the other pigments in the leaves—which may be red, yellow, or gold—can be seen, giving the leaves their bright fall colors. Later, a signal is given by the tree to the cells at the base of each leaf stem, or petiole, to die, and eventually the leaves fall to the ground.

Another characteristic sight in the northern hardwood forests of Vermont is the collection of sap from sugar-maple trees to make maple syrup. In the late winter, when days become warm while nights are still cold, sugar-rich sap begins to move inside the maple tree just underneath its bark to transport stored sugars from the roots of the tree out to the buds to make new leaves in the spring. Indians were the first to learn how to tap into a sugar maple and collect the sap, and they later taught this skill to European colonists. The colonists then discovered that they could boil the sap after it was collected to drive off much of the water and increase its sugar concentration. By this process, thirty to forty gallons of sap are used to make one gallon of maple syrup.

Many plants are commonly found in the understory and herb layers of northern hardwood forests, including evergreen woodfern, Christmas fern, red trillium, white wood aster, starflower, and, at higher elevations, hobblebush. These species grow best where conditions are warmer and the ground less covered by snow throughout the year than is the case at high elevations.

Transition Hardwood Forests

A number of forest-community types are found in the transition areas between northern hardwood forests and hardwood forests characteristic of warmer, more southern latitudes. Northern hardwood species, especially sugar maple, American beech, yellow birch, and white ash, may still be dominant, but species better adapted to conditions warmer and drier than those found in most of Vermont today are increasingly abundant. These transition forests are found predominantly in the Champlain Valley, the

Valley of Vermont, the lower slopes of the Taconic Mountains, and the southern portion of the Piedmont. In some locations, seasonal drought and poor soils are dominant forces acting on the growth and abundance of the trees in transitional forests.

As with both the cold-climate and northern hardwood forests, a number of variants of the transition hardwood forests are recognized, depending on the dominant species. All of them are considered rare to uncommon in Vermont, in some cases due to their restricted habitat requirements and in others due to elimination by European settlers or by development during the twentieth century. All together, transition forests probably comprise less than 5 percent of the total forestland in the state.

Along the shores of Lake Champlain and its islands where the soil is shallow and built on limestone, the lake bluff cedar-pine forest is present (see fig. 7.5). Here, a number of low-elevation conifers—particularly northern white cedar, red pine, white pine, and hemlock—are found along with sugar maple, shagbark hickory, and white ash. The herbs associated with this community are primarily sedges and a number of fern species.

In the lowlands of the Champlain Valley and in the Taconic Mountains is found an oak-hickory–northern hardwood forest. This community type represents the transition between the beech-birch-maple forest and the oak-hickory forests of the south, and is dominated here by red oak, white oak, shagbark hickory, beech, paper birch, and sugar maple. Where conditions are particularly dry, such as on hilltops in the Champlain lowlands and Taconic Mountains, eastern hop hornbeam may be abundant. Where conditions are wetter and more prone to flooding, trees more common in swamps (see chapter 9) are present, particularly swamp white oak. Woodlands that develop on and below rockfalls at the bases of major cliffs in the warmer parts of the state have a greater abundance of basswood, hackberry, and butternut, along with the shrub bladdernut. These transition hardwoods-talus woodlands are like the other talus woodlands throughout the state: sparse, with small trees and shrubs.

Where deltas formed at the edges of the retreating glaciers more than 12,000 years ago, the soil is sandy, and a particular fire-adapted community of pitch pine, white pine, oak, and low heath shrubs such as blueberry and huckleberry develops. Although it was probably once much more widespread in the deltas of the Winooski, Lamoille, and Missisquoi Rivers, this pine-oak-heath sandplain forest community has become restricted over the last two hundred years to only a few small sites due to fire suppression, human development, and extraction of sand. Most remnant sites are found in the northern lowlands of the Champlain Valley, but a few sites exist in the southern Piedmont along the Connecticut River.

A number of other community types form on rocky ridges and cliffs

FIGURE 7.5. A lake-bluff cedar-pine forest community. Photo by Stephen Trombulak.

where the soil is so thin and moisture so lacking that only an open woodland can develop. These include red-pine woodlands on rocky ridge tops and on slopes where soils are dry and fires are frequent, pitch pine–oak-heath rocky-summit woodlands on dry acidic ridges, dry oak woodlands on steep, well-drained south-facing upper slopes, and red-cedar woodlands on dry cliff tops.

Because transition hardwood-forest communities are so diverse, few herbs and shrubs are specific to them, but they all tend to be best adapted to warm, dry climates and the characteristic kinds of soil—for example, shallow calcareous soil, deep clay soil, and acidic sandy soil.

Forest Animals

A number of animals are, or were, associated primarily with forest communities in Vermont. An awareness of them is important for understanding forest communities, how both Native Americans and European settlers survived in Vermont, how human actions have altered the forest fauna over the last 250 years, and the true extent to which forest communities have returned in the last hundred years. None of these animals is unique to Vermont, and most are found widely in forest communities elsewhere in North America.

Of the more than 250 native species of birds in Vermont, almost half are found predominantly in forests and woodlands. Most of the perching birds —a large group of birds that includes the thrushes, warblers, and sparrows —are forest dwellers. The warblers are a particularly beautiful part of Vermont's avifauna; many of the thirty or so species of warblers are brightly colored with golds, yellows, blues, and greens. Males sing melodiously in the spring to advertise the locations of their territories and to attract mates. Warblers, like woodpeckers, vireos, flycatchers, nuthatches, creepers, wrens, kinglets, thrushes, and many other forest-dwelling birds, feed primarily on insects and other small invertebrates, using their long, thin beaks in a variety of ways to capture their prey.

The ten or more species of sparrows, finches, cardinals, and grosbeaks that live in Vermont's forests are primarily seed eaters. Their short, thick beaks are adapted specifically to crushing seeds and nuts. Some seed eaters have extreme adaptations for foraging. Red crossbills and white-winged crossbills have beaks whose tips cross, giving them the plierslike leverage they need to pry open cones and reach the seeds inside. Many other forest birds, such as black-capped chickadees and blue jays, are opportunistic feeders, eating both seeds and insects as they are available through the seasons.

Certain birds are associated with particular forest-community types. Some species, such as blackpoll warbler and pine siskin, are closely associated with high-elevation spruce-fir forests, whereas others, such as white-breasted nuthatch, red-eyed vireo, and chestnut-sided warbler, much prefer hardwoods. Birds will even select forests based on the size of the trees. The wood duck, which nests in tree cavities, prefers red maple among swamp hardwoods, but only those greater than sixteen inches in diameter. Alder flycatchers, on the other hand, are found where the forest comprises predominantly small seedlings.

Some of the largest forest birds eat a mixed diet of seeds, nuts, mushrooms, and insects. Ruffed grouse, spruce grouse, and wild turkey are stout-bodied ground-dwellers that rarely fly very long distances, preferring to walk through forest and field during the day in search of food. Ruffed grouse are common throughout Vermont. As a potential predator gets close to these well-camouflaged birds, the grouse fly up in a short burst of flapping wings. In the spring, males attract females not by song but by making low-pitched drumming sounds with their wings. Spruce grouse are much less common, being restricted in Vermont to only a few locations in the spruce-fir forests of the Northeastern Highlands.

The wild turkey is the largest ground-dwelling bird in North America. Once common throughout Vermont, it was extirpated by the mid-1800s through a combination of forest clearing and overhunting. As forests became widely reestablished in Vermont, turkeys could once again thrive. In 1969 and 1970, the state relocated thirty-one turkeys from western New York to the Taconic Mountains, where their numbers grew. Subsequent relocations of turkeys to other parts of Vermont have resulted in healthy populations, particularly in the southern half of the state.

Birds of prey are also numerous in Vermont's forests. Day-active hunters such as the sharp-shinned hawk, Cooper's hawk, and goshawk are found throughout Vermont, as are the night-active owls, including the great horned owl, barred owl, and eastern screech owl. These predators all have strong, grasping claws and short, hooked beaks for killing small mammals and birds.

Many of the birds found in Vermont's forests are migratory, spending only part of the year here. Most of them migrate southward during the winter to islands in the Caribbean Sea or to Central and South America. They might well be considered tropical species that come northward in the spring and summer to breed where summer days provide many hours of daylight and bountiful insects for feeding the young chicks.

A few species—such as the American tree sparrow and much more rarely the northern hawk owl—breed further north and come south in the winter. Still other species, such as the wild turkey, forest hawks, woodpeckers, and northern cardinal, are present throughout the year, breeding in the spring and summer and foraging as best as they can through the winter.

Not surprisingly, most of the mammals in Vermont are associated primarily with its forests. Of the fifty-four mammal species native to Vermont that are still present, eighteen are rodents. Many of these—like the deer mouse, white-footed mouse, red-backed vole, and flying squirrel—are small, secretive, night-active creatures that make their livings feeding on seeds, fruits, leaves, and occasionally insects. A few, such as the gray squirrel, red squirrel, and eastern chipmunk, are active during the day.

Almost all mammals have specific habitat preferences that make them more common in one forest type than another. Red-backed voles prefer spruce-fir forests, especially where fallen trees provide easy runways through the understory. Many more species, however, are found more commonly in beech-birch-maple forests, including the eastern chipmunk, southern flying squirrel, deer mouse, woodland jumping mouse, and hairy-tailed mole. The greater diversity of mammals in the hardwood forests might be due simply to the milder climate there.

One forest-dwelling rodent of special interest is the porcupine. Porcupines are large animals, weighing as much as twenty-five pounds, and are easy to recognize by the dense pelt of sharp quills over their body. Quills are modified hairs that, because their tips are covered with recurved barbs, easily become imbedded into anything they strike. Porcupines like to eat cambium, the layer of growing tissue underneath the bark of a tree. Most trees can survive a little bit of foraging by a porcupine, but if the porcupine eats all the way around the trunk, the tree will eventually die. When fishers, weasel-like animals about the size of small dogs and the primary predator of porcupines, were extirpated in Vermont by trapping in the mid-1800s, the population of porcupines grew so large that some foresters felt that many of the remaining forests were threatened. Fishers were eventually reintroduced to Vermont in the 1960s specifically to control the porcupines and help protect the forests.

A number of other small mammals, including several species of shrew, mole, and bat, also live in forests. Shrews and moles are primarily carnivorous, feeding on insects, worms, slugs, spiders, centipedes, millipedes, and mice. They are voracious, eating more than three times their own body weight every day and never stopping to rest for more than an hour and a half at a time before going off again in search of more food. Most shrews in Vermont, like the masked shrew and the short-tailed shrew, prefer to be active under the leaf litter on the forest floor.

Moles hunt primarily underground. With their broad, flat paws they dig shallow tunnels where they hunt for invertebrates that live in the soil. The most unique of the moles is the star-nosed mole, which has a ring of twenty-two pink, fleshy tentacles that surrounds its snout. These tentacles are sensitive to touch and help the star-nosed mole to find food since its underground environment is always dark.

Nine different species of bat are found in Vermont during the summer. All of them make their homes inside of or under the bark of trees near water, where, during the night, they feed on flying insects. Since the colonization of the region by Europeans and the building of houses and barns, some species of bat have taken readily to living in these human-made structures as well.

Six species of bat hibernate during the winter in caves and abandoned mines in both the Taconic and the Green Mountains, as well as in a few abandoned mines near the Connecticut River. The winter populations of bats in Vermont have declined dramatically since the 1960s, mirroring the decline in bats throughout North America. Hibernacula that once had winter populations numbering in the thousands have had only a few hundred in recent years; those that numbered in the hundreds have had fewer than ten. The largest winter populations are now found in abandoned talc and copper mines, habitats that have become available to bats only in the last hundred years or so. Declines in bats are thought to be attributable to a combination of the increased use of pesticides—which poison insect-eating animals like bats—as well as disturbance to bats during the winter by people who enter hibernacula.

The snowshoe hare is also a resident of Vermont's forests. This animal prefers to browse on the growing tips of shrubs and small trees; in high numbers, hares can have a powerful effect on the density of understory vegetation. Snowshoe hares change the color of their fur from brown in the summer to almost pure white in the winter, allowing them to hide from predators.

Most of the attention paid to forest mammals by people is to carnivores and large herbivores. The forests of Vermont are, or were, home to a large number of carnivore species that include the timber wolf, coyote, foxes, black bear, raccoon, weasels, fisher, marten, striped skunk, wolverine, mountain lion, bobcat, and lynx. Because of a cultural bias of European settlers against predators, the clearing of forested habitat, and, at one time, an economic value for pelts, virtually all mammalian carnivores in Vermont were affected by human colonization of this area. As mentioned above, fishers were trapped out of the state and later reintroduced. Similarly, pine martens were trapped out by the 1940s; a reintroduction of 115 animals was carried out in the Green Mountains from 1989 to 1991, but whether it will ultimately be successful is still unclear. Lynx have also been reduced to such a level that they are, in effect, ecologically extinct from the state and seen only rarely.

Black bears are today the largest carnivores in Vermont, weighing as much as four hundred pounds. Although hunted to very low numbers in the 1800s, they are now widely distributed in larger tracts of woodlands throughout the state but are wary of people and not often seen. They live in a variety of forest communities, especially near water. Even though they are classified as carnivores, they are quite omnivorous and even depend on quantities of beechnuts to put on enough fat for winter hibernation. Large stands of beech trees are critical resources for bear populations.

Two carnivores were completely extirpated from the region, and although still absent, are topics of great discussion in Vermont today. Mountain lions, known by many in Vermont as catamounts (derived from the term "cat of the mountains") were once found throughout North America,

FIGURE 7.6. A timber wolf. Reproduced with permission from Jim Brandenburg and Minden Pictures.

and since the great mammalian extinction ten thousand years ago have been one of the largest predators on the continent south of the Arctic, second only to the grizzly bear in the West. Because of hunting by humans and the transformation of the Vermont landscape (described in chapters 4 and 5), the population of mountain lions in Vermont was reduced to the point where it could no longer sustain itself; the last native mountain lion known for certain in Vermont was shot in 1881.

Although numerous people have reported sightings of large catlike animals over the past several decades, no well-documented evidence of mountain lions in Vermont was found until 1994, when numerous scats of a mountain lion were found in the northern Piedmont near Craftsbury Common. Whether this individual is a long-distance migrant from a population in more northern Quebec, an escapee from captivity, or a part of a larger but secretive population that remained in Vermont after the 1880s is not yet known.

The animal that probably most excites the imagination of Vermonters is the timber wolf (see fig. 7.6). This highly social animal was once distributed widely throughout North America as one of the top-level carnivores, preferring in the Vermont region to prey on deer and beaver. In areas where human settlement largely eliminated those species, wolves switched to preying on livestock like sheep and cattle, which led to increased conflicts with humans and eventually to their extirpation throughout most of their range. Wolves were eliminated from Vermont by the end of the nineteenth century, and up until the early 1990s wolves were found in North America only in Alaska, Canada, and the northern border states of Minnesota, Wisconsin, and Michigan. Through the efforts of the U.S. Fish and Wildlife Service, the timber wolf was reintroduced to central Idaho and Yellowstone National Park in Wyoming in 1995, and populations now seem much more secure in the north-central states. In light of this success, an explosion in the abundance of beaver and deer in the northeastern United States has many people advocating the reintroduction of wolves, particularly in northern Maine or Adirondack Park in New York—a reflection of changing attitudes toward carnivores. Successful reintroductions in either place would probably result eventually in wolves dispersing into the Northeastern Highlands and Green Mountains of Vermont, so this species might one day again be part of our forest fauna. Indeed, confirmed sightings of wolves in Maine in both 1993 and 1996 indicate that wolves might stage a comeback on their own, moving in from more northern populations.

The other large mammals of Vermont's forests are deer and moose. Deer were probably quite numerous throughout the region prior to European colonization, but, as described earlier, hunters practically eliminated them from Vermont by the end of the 1700s. Because they were so important as a source of meat, deer were reintroduced to Vermont from New York in 1878. However, the environment into which those deer were reintroduced had changed drastically in the preceding hundred years. The deer's predators, primarily wolves and mountain lions, had been eliminated, and much of the forest had been cleared, which created a vast amount of the open space and edge habitat preferred by deer during the summer months. As a result, deer populations in some locations in Vermont exploded, as they did throughout eastern North America, and their populations today are probably far in excess of what is ecologically sustainable.

The other large herbivore that has returned to Vermont is the moose, one of the largest mammals in North America, standing six feet high at the shoulder, nearly twice the height of a white-tailed deer. Once one of the prime prey of the Abenaki Indians, the moose was eliminated from Vermont by hunting and land clearing in the 1800s but began to reestablish itself in the 1960s in the cold-climate forests of the Northeastern Highlands from populations in Canada. Although they are primarily a boreal forest species, their population has continued to expand southward through Vermont. They are seen occasionally in northern hardwood forests even in the lowlands of the Champlain Valley.

Another group of vertebrates found in Vermont is the reptiles, animals that, with few exceptions, have scaly skin and lay shelled eggs. Of the eleven snake species in Vermont, the most common is the garter snake. This gentle yellow and dark green reptile is often seen sunning itself along the sides of roads and trails throughout the state. It preys on small animals, such as insects and frogs, and therefore is an important part of the forest food web. The garter snake is found in a wide variety of habitat types but is especially common in northern hardwood communities near wetlands. It is well adapted to life at northern latitudes. Rather than lay eggs like most snakes, the garter snake gives birth to live young. This allows the snake to reproduce successfully even where the ground is too cold through most of the year to permit eggs to hatch. Thus, garter snakes are found even in far northern reaches of Canada. Like many other snakes, garters survive the cold winters typical in most of their range by crawling deep underground, below the level to which the ground freezes, and curling up in a large ball that can include several hundred other snakes. With the spread of houses in Vermont, the crevasses between the ground and a house's foundation have become popular hibernation sites. In the spring when the snow melts and the ground warms up again, a homeowner can discover snakes crawling up from the base of the house and lying in the yard for a few days before they disperse for the summer.

Most other species of snake in Vermont can be found to some extent in forest or woodland habitats. All are much less common than the garter snake, and their distributions within the state are poorly documented. The eastern ribbon snake, rat snake, and brown snake seem to have all or most of their geographic distribution within Vermont in the Champlain Valley. In Vermont, all of these snakes are near the northern end of their overall geographic distribution, and it is not surprising that they would be located primarily in the warmest biophysical region in the area. However, in many parts of the state, particularly the Northeastern Highlands and the Taconic Mountains, studies of reptiles and amphibians have not been extensive, and much new information about these animals will be discovered as more people learn to enjoy searching for them.

Of all the snakes in northern New England and the Maritime Provinces, only one is venomous. Using hollow, needlelike fangs, the timber rattlesnake injects a poison into its prey or a larger animal that threatens it. A rattlesnake is easy to identify: Its head is nearly triangular, and it has a rattle made of tough, dried skin at the end of its tail. No other snake in this area has these characteristics. This animal, although once distributed much more widely throughout Vermont, has been reduced to only a few populations in forests near cliffs and talus slopes in the west-central part of the state. At first they were hunted for bounty and simply because people feared them, but in recent decades, even as the populations were reduced to levels that made accidental interactions with humans extremely rare and even when they were protected legally, they continued to be collected illegally by a few people for the pet trade. Now they are all but extirpated from the state.

Amphibians are a group of vertebrates that, as their name implies, in general live part of their lives in water and part on land. Since most amphibians in Vermont lay their eggs in a body of water of some kind, they will be described as part of the state's aquatic communities (see chapter 9). One species, however, the redback salamander, lives its entire life on land and lays its eggs in moist places inside decomposing logs and stumps. It is about four inches long, usually with a dull red stripe down its back, set off from its dark brown sides. Like more than half of all the salamander species found in Vermont, the redback salamander does not have any lungs; it gets all of its oxygen through its skin, which is why it must always keep itself moist. The redback salamander is one of the most common vertebrate species in New England and can usually be found by turning over rocks and logs in most northern hardwood forests. Redback salamanders cannot live in areas where the soil is too acidic, and therefore this species is an early-warning system for the acidification of forest soils (see below).

The eastern newt has an adult stage that is entirely aquatic. The female lays her eggs in the shallow water of small ponds, where they hatch into juveniles that are restricted to the water. These juveniles, however, eventually metamorphose into terrestrial juveniles, called red efts. These small salamanders, marked by indistinct red spots on an orange-red background, wander fearlessly through the leaf litter of forests throughout Vermont, foraging for small invertebrates. The bright red color of the efts is a signal to potential predators that they are quite toxic; the skin of a single eft contains enough poison to kill a medium-sized bird. The efts live for three to eight years in the forest before they migrate to a pond and metamorphose into fully aquatic adults.

Although vertebrate animals are the best known among the forest residents, they are by no means the most numerous. Thousands of species of spiders, butterflies, ants, beetles, millipedes, centipedes, flies, moths, bees, worms, and a host of other animals are found here. Some species are well known, such as many of the moths, butterflies, ground beetles, and crop pests. Unlike for the birds, mammals, reptiles, and amphibians, however, a full accounting of these species in Vermont has never been made, and knowledge of the terrestrial invertebrate fauna is poor. It is not even known to the nearest thousand how many invertebrate species actually live here. Better study of the natural history of these groups remains one of the greatest natural-history challenges for the future.

All the organisms that live in an area contribute to the area's biodiversity: the variety of living organisms considered at all levels of biological organization, including genes, species, and communities. All forms of variation in the natural world—from variation in the genetic information that creates differences between individuals in a species to the variation in communities across a landscape—contribute to biodiversity. Much of this variation, however, is not easy to see when simply looking at nature, and often "biodiversity" is used to refer, although incompletely, to the list of species that are found in an area.

The return of the forests through the twentieth century allowed many forest species to return or be reintroduced successfully. Many species, however, have not returned, so Vermont forest communities today are different from those at the time of European settlement. In addition to timber wolves and mountain lions, four other species were eliminated. Wolverines, large, fierce carnivores related to weasels, were extirpated from Vermont by the mid-1700s. Although they were probably never numerous, owing to their solitary natures and home ranges of sixty square miles or more, they were important top-level carnivores. Both elk and caribou—large hoofed herbivores like deer and moose—were eliminated from Vermont. Although attempts have been made to reintroduce both of them into other parts of New England from more northern populations, none has been successful. Passenger pigeons, herbivorous birds whose flocks were known to have included millions of birds, were extensively trapped for food throughout eastern North America in the late nineteenth century. Huge numbers of them were shipped on ice south to New York City from Plattsburgh, having been trapped throughout Vermont and eastern New York. The species was extirpated from northern New England by the late nineteenth century and driven to complete extinction in 1914.

Forest Processes

A community is more than just a collection of species that live together in a local area; species also interact with each other and with their environment. These ecosystem processes weave species and all other components of biodiversity together into an integrated web. While a single species may possibly arrive into or disappear from a community without any noticeable effect on how the community functions, species are not all interchangeable, and the connections among species are important to an understanding of the community. Knowledge of forest processes is critical for understanding the effect human settlements have had on forest communities and for shaping future conservation efforts.

Many kinds of processes operate within the forests of Vermont. The first is that of the flow of energy into a community. The ultimate source of energy for almost all life on Earth is the sun. Sunlight reaches the earth's surface and, in most cases, is absorbed and warms up the air, water, or ground. Some of the wavelengths of sunlight are captured by plants, which through the process of photosynthesis transform solar energy into chemical energy. Photosynthesis is the creation of sugar molecules from carbon dioxide and water using energy from the sun, with oxygen released as a by-product. The bonds between the atoms of a sugar molecule are essentially stored energy; the more sugar molecules produced by photosynthesizing plants, the more energy that can be transported throughout the community. All other things being equal, the more energy there is, the more individuals of each species that can be supported and the more different kinds of species that can live in the community.

The amount of solar energy transformed into chemical energy by the plants in a community is called its primary productivity. The primary productivity of a community is based on a number of things, including how many plants are present, the length of the growing season, soil fertility, and how much water is available. In general, deciduous forests in middle latitudes such as Vermont have a level of primary productivity equivalent to about 10,000 pounds of dry weight of plant matter per acre per year, a little above average compared to different community types around the world. Tropical rain forests, because they have longer growing seasons and more water, have productivity levels of about 15,000 pounds of dry weight plant matter per acre per year. Deserts and tundra have productivities less than a tenth of that.

An important by-product of photosynthesis is free oxygen, a molecule on which almost all life depends. It is thought that a substantial fraction of all of the oxygen used by living things is produced by the forests of the world. Without its production, life on Earth would eventually cease.

Another important consequence of photosynthesis is the removal of carbon dioxide from the atmosphere. This molecule is released when sugars and other carbon-rich molecules are burned, such as in a fire or in an automobile engine. Carbon dioxide is also produced by living organisms as they respire: Oxygen is inhaled and carbon dioxide is exhaled. Carbon dioxide is one of the "greenhouse" gases, which are capable of trapping heat in the atmosphere and whose atmospheric concentration is increasing and which have been implicated in the rise in global temperatures. The regeneration of forests around the world, such as has occurred in Vermont over the last hundred years, would remove from the air a substantial fraction of the carbon dioxide that is produced from burning fossil fuels. Our ability to control the climate warming that will result from our culture's production of carbon dioxide might depend in part upon the regrowth of the world's forests. Vermont offers a model for such widescale reforestation, but the model is somewhat imperfect. As we argued in chapter 6, Vermont's forests were able to return largely because Vermont's people import most of their resources. Vermont's reforestation came at the expense of deforestation elsewhere.

Another set of interactions to consider is the kind that takes place between the species themselves. Every species has some kind of relationship with others in the community: as competitors for food and space, as predators and prey, as parasites and hosts. These kinds of interspecific interactions are influential in determining whether a species can persist in an area and how its population will behave. The connections among species defined by who eats whom are collectively called a food web. Most food webs are structured so that chemical energy that has entered the community by the photosynthesis of plants passes through the web from one link to the next—plants to herbivores to carnivores—in such a way that a wide diversity of species is supported.

The importance of food webs in forest communities is understood by looking at a simple example of what can happen when interactions among species in a web are altered. With the disappearance of timber wolves and mountain lions, the original top-level carnivores in this region, there were no checks on the populations of their herbivore prey, particularly moose, white-tailed deer, and beaver. All three of these species themselves were also seriously reduced or extirpated, but because of their popularity as game species they were reintroduced or allowed to become reestablished. These reintroductions took place, however, into communities where the predators had been removed. As a result, their populations increased so much that they have altered the composition and physical structure of many of the natural communities where they live.

Species also interact with each other by storing, transporting, and transforming the nutrients on which all species depend. Many different elements are essential to plants and animals, including nitrogen, carbon, calcium, potassium, phosphorus, magnesium, sulfur, and at least a dozen others. These nutrients help all living things build parts of their bodies, transform and transport energy, and regulate metabolism. Each of these nutrient elements has its own unique cycle or pattern of movement and transformation in the environment. During its cycle, an element can be combined into different molecules by different processes and can be stored in many different places.

One of the most important nutrient cycles in any community is that of nitrogen, a fundamental component of many different molecules in an organism's body, especially of protein. Without enough nitrogen, for example, plants grow poorly and look spindly and weak. The nitrogen cycle demonstrates the wide range of interactions that exist among organisms. When organic matter decomposes on the forest floor, bacteria and fungi in the soil break up these nitrogen-rich molecules and release much of the nitrogen into the soil as ammonia. However, in order to be usable by most living organisms, ammonia must first be transformed by bacteria into other nitrogen molecules, such as nitrite and nitrate. Nitrogen can also be made available to plants from the nitrogen in the atmosphere after first being transformed into nitrite and nitrate by specialized nitrogen-fixing bacteria. Some bacteria form symbiotic associations with certain species of plants, and live in special nodules on the roots of the plants.

Clearly, the entire health of the forest depends on the presence and action of many seemingly insignificant organisms and the maintenance of soil conditions that allow these plant species to persist.

Disturbance is another important process in forest communities. The natural world is full of significant disturbances, including fire, wind- and ice storms, disease, insect and fungal attack, and flooding. Under most conditions, these processes do not work in opposition to the health of the forest. In fact, the rich diversity of species found in many of these communities is dependent on such disturbances taking place. Disturbances in forests open up gaps in the canopy, allowing sunlight to reach the forest floor. These gaps give herbs and shade-intolerant trees the opportunity to grow and produce seeds. The deaths of trees in an area, if they are allowed to fall and decompose, return nutrients to the soil, which can then be used by other species. Some animals even depend on disturbance for survival. Woodpeckers, salamanders, and many species of insects depend at some point in their lives on dead trees for food and shelter.

Every forest community has its own natural scale and frequency of disturbance to which species are adapted. Because the climate in Vermont during the summer tends to be wet, fires tend to be small and infrequent. This differs from western North America, where hot, dry summers usually result in fires that spread over many thousands of acres.

Windstorms, including hurricanes, occasionally create very large disturbances in Vermont's forests. For example, in 1938 a hurricane swept up the Green Mountains from the south, creating numerous gaps in the forest several acres in size. Storms like this occur on average once every fifty years or so in southern New England but have occurred only twice in Vermont in the last two hundred years. In January 1998, an ice storm hit the northern half of Vermont, killing or otherwise damaging trees over many hundreds of thousands of acres. An ice storm of this size has never before been recorded in Vermont, and it is expected to have a dramatic long-term effect on its forest communities. On a more modest level, wind can cause death and rebirth in small spruce-fir stands in the Green Mountains (see fig. 7.7). At higher elevations, where trees are exposed to steady wind throughout the winter, blowdowns are common.

FIGURE 7.7. Balsam fir knocked down by wind at high elevations in the Green Mountains. Photo by Stephen Trombulak.

Because nutrient cycles involve so many species, changes in community structure and the health of the soil can have large consequences on whether nutrients are kept within the community or eventually lost. In general, few nutrients are lost from forests that are not disturbed. Most nutrient loss from forest communities takes place following disturbances, such as fire, tree death, road building, development, logging, or changes in soil chemistry by acid rain. Forest recovery from these disturbances is affected strongly by the amount of nutrients removed and by how seriously the disturbance affects the species that are important in the cycles, such as soil bacteria. The widespread production of potash from the trees felled during the early days of European colonization was particularly hard on nutrient levels since the potash was shipped away. Agriculture of all types has the same effect; nutrients are transported from the region as crops or animal products and need to be replenished in the form of fertilizers.

Many other types of disturbance occur in forest communities, such as individual tree falls and rock slides. Rock slides at the bases of cliffs can occur so regularly that specific talus communities form there (see tables 7.1 and 8.1), comprising species that are adapted to the frequent disturbance.

Disturbance in forest communities creates the conditions necessary for regeneration of the same species or recolonization of the area by other species. Many plants found in forests in Vermont are actually better adapted to open habitat (see chapter 8), including early successional species that are the first to grow in forest openings. If all natural disturbances were somehow prevented, many species native to forest communities in Vermont would probably disappear.

With all of these processes causing near-constant change in forest communities, we are left with the question of whether forests can be considered ecologically stable. The answer to this depends very much on how stability is defined. If stability implies constant, unchanging conditions, then forests are not stable. Windstorms knock down trees frequently and unpredictably, creating patches of early successional plants. Spruce budworm, a native insect, periodically increases in number and kills spruce and balsam fir. Disease can sweep through an area and remove a single susceptible species like chestnut or elm, both of which were largely eliminated by diseases introduced in the last hundred years. Over longer periods of time, glaciers sweep down from the north and scrape the ground bare of all life.

However, if stability implies the existence of checks and balances, where under natural conditions the abundance and distribution of most species recovers after a disturbance, then forests are somewhat stable. At least between glacial periods, species of predators and prey maintain a balance in number and succession leads to regular patterns of species replacement following disturbance. As forest ecologist Tom Wessels noted, "Disturbance is a force that counters the successional growth of ecosystems. Together, disturbance and succession are like the opposing but balancing yin and yang of Taoist thought." Forests exist as shifting mosaics of older and younger stands distributed across a landscape. Forest communities are dynamic, and within geological periods operate in more or less predictable ways.

Trends in Forest Health

We have said much already about the "natural" conditions of the forest, by which we mean the conditions that would exist if humans, either Indians or

Europeans, were not present. In this context, "natural" is contrasted with "cultural," not with "unnatural." Humans are clearly a part of nature, being a species that needs to derive its ultimate resources for survival from the rest of nature. Yet it is useful to make a distinction between conditions that exist as a result of nature operating in its own way in its own time and conditions that are the result mainly of the cultural activities of humans. Although the development of human cultures in Vermont has altered natural conditions in a wide variety of ways, four changes in particular are of note because of the magnitude of their impact on forest communities.

The first is the change in forest cover over time. As noted, European colonists reduced the amount of forest cover in Vermont from about 95 percent to something less than 40 percent by the late 1880s. Forest clearing in Vermont was so extensive and the native flora and fauna prior to the clearing were documented so poorly that the consequences of the clearing will never be known completely. Based on the writings of some early Vermonters, such as George Perkins Marsh, and on what is happening today in other regions undergoing deforestation, it is certain that clearing resulted in a tremendous increase in soil erosion. Many plants and animals that depended on cover and soils provided by closed forests were also probably lost during this time. The biological legacy that was lost during the clearing will never be known fully.

Another transformation that occurred was a change in the ages of the trees in the forest. Most people think that the typical Vermont forest of today—numerous small trees with a dense understory of shrubs—resembles a Vermont forest of the past. In fact, the forests created over the millennia by the native plants and animals of this region and the forests colonized by both Indians and Europeans were drastically different from what is seen today. Prior to colonization by Europeans, Vermont was dominated by forest communities that can best be called old-growth or ancient forests. It is difficult to define precisely what an old-growth forest is for any particular community type, but in general old growth is characterized by a sizable percentage of trees that are of advanced age for the species (some species, like hemlock, naturally live for hundreds of years, and others, like balsam fir, rarely live past eighty years), downed logs, standing snags, uneven-aged structure of trees that form the forest canopy, presence of gaps caused by tree falls, pit-and-mound topography, and uncompacted soils. Identification of old growth requires good knowledge of the biology of the species present and careful attention to the physical structure of the community and the soil.

It is likely that prior to colonization by Europeans 75 percent of Vermont's forests were in old-growth condition at any one time. In an area that probably comprised over 5.5 million acres of forest land (out of Vermont's roughly 6 million acres total), that indicates the continuous presence of over 4 million acres of old growth. The clearing of forestland that occurred during the last two hundred years resulted in a tremendous loss of these ancient forests. Today, fewer than two thousand acres of old growth have been identified in Vermont. These are scattered widely through the state in numerous small stands, mostly less than one hundred acres each. Particularly noteworthy are Lord's Hill Natural Area in Groton State Forest, which includes about thirteen acres of old-growth beech-birch-maple forest, a stand of approximately 270 acres of red spruce within White Rocks National Recreation Area in the Green Mountain National Forest, and the Battell Biological Preserve (owned by Middlebury College), approximately two hundred acres of hemlock forest, with some trees more than three hundred years old.

Although many parts of the state, particularly in the Taconic Mountains, have not been surveyed adequately for the presence of stands of old growth, it is unlikely that very many acres have yet to be discovered. Therefore, it seems reasonable to assume that nearly all of Vermont's old growth has been eliminated. This is profoundly unfortunate for at least two reasons. First, studies elsewhere in New England have shown that a variety of species, particularly insects and lichens, are associated primarily with old growth. The huge reduction of old-aged forests in Vermont almost certainly means that many species have been lost from the state due to the elimination of their habitat.

Second, the absence of old growth radically transforms human perceptions of a natural forest. The stands of old growth that exist today provide the only glimpse of what the pre-European settlement forests in Vermont might actually have been like. Yet they are so small in area and scattered so widely that few people have actually seen one, and those old-growth tracts that remain are almost certainly atypical.

Further altering forest communities are the species that have been brought into the area by humans. Exotic species have been introduced purposely into North America for a number of reasons—primarily economic development and nostalgia—but most of introductions have occurred by accident, invaders being brought here as passive travelers on other products, such as food or timber. Exotic forest species of all types are now widespread in Vermont, especially plants, invertebrates, fungi, and bacteria. Of the trees now found in Vermont's forests, for example, several were introduced here by European colonists, including Norway spruce, Austrian pine, Scots pine, crack willow, white poplar, Lombardy poplar, and European buckthorn. In all, fifty species of plants, from herbs to trees, out of a total of about 660 found in forest communities in Vermont today, were introduced into the forests of this region by humans in the last four hundred years, although generally they have not spread far from where they were planted.

Most of the health-related problems that now face particular tree species in Vermont were also introduced from elsewhere. Two of the most widespread are chestnut blight and Dutch elm disease. Chestnut blight is caused by a fungus that was introduced accidentally in New York City at the turn of the century. It was apparently a passive traveler on stocks of Asiatic chestnuts imported for the horticultural industry. From there, it spread rapidly throughout the range of the American chestnut. This tree was once a dominant or co-dominant species throughout southern New England and the mid-Atlantic region of eastern North America. In less than fity years after the blight was introduced, however, chestnuts throughout their range were dead or dying, and the species now remains as only a few isolated trees or as small sprouts at the bases of old stumps. The oak-chestnut community so common one hundred years ago has been transformed into a mixed oak community. Similarly, Dutch elm disease is the result of a fungus that was first noticed on this continent in Ohio in 1930 and spread from there throughout eastern North America. Although small elms still persist in the understory in many Vermont forests, they rarely reach their former heights of more than one hundred feet, and have been eliminated as a dominant species in the Northeast.

Other introduced species that are transforming the forests of the region include gypsy moth, beech-bark scale, pear thrips, red-pine adelgid, hemlock woolly adelgid, balsam woolly aphid, European pine sawfly, white-pine blister rust, and red-pine scale. All of these are responsible to one level or another for decreasing the life spans of individual trees, increasing the rate of disturbance in forests, and changing the patterns of species abundance and distribution.

Finally, recent technological developments have significantly changed the quality of the air. Of greatest concern with respect to forest health has been the increase in acid precipitation throughout the region. Whenever gasoline or coal is burned, nitrous oxides and sulfur dioxides are released into the air, where they combine with water to make nitric acid and sulfuric acid. Rain, snow, and fog deposit these acids onto vegetation and the ground, changing soil chemistry and sometimes causing physical damage to plants. In the 1980s, it was noticed that acid precipitation in the northeastern United States was extensive and was caused largely by coal-burning power plants in the midwestern United States and in Canada. At the same time, red spruce in the Green Mountains, as well as in the Adirondack and White Mountains, were dying at a much greater rate than before, accompanied by noticeable damage to their needles. It has been assumed that the increase in red-spruce death has been the result of acid precipitation, but whether this is true—and if so exactly why the spruce are dying and other species not—is still unknown. Acid precipitation has been implicated recently in the leaching of nutrients, particularly calcium and magnesium, from forest soils in New Hampshire; the consequence of this change in soil quality is that for at least the past twelve years those forests have completely stopped growing, which has serious implications for the ability of the forests to regenerate after disturbance.

Ground-level ozone has also increased in Vermont. This gas, also a by-product of the burning of fossil fuels, is toxic to plant tissue and has increased the mortality of trees and other plants in the region quite dramatically. Many other pollutants that have also increased in recent years, such as benzene and formaldehyde, are found in concentrations far in excess of what is considered healthy for humans and many other animals.

Declines in canopy cover by sugar maple, white ash, and red oak have been recorded in this region in recent decades. These declines come probably not as a result of any one factor but from multiple assaults: air pollution, disease, changes in community structure, and nutrient depletion. They might be the earliest indications of more dramatic changes yet to come in Vermont's forests.

Forests have been the dominant natural-community type in this region for the last 12,000 years and have provided much of the motivation for human occupancy since people first migrated here. They comprise many different species, connected to each other by complex interactions and processes. Many of these species declined or were eliminated by forces set into motion by European colonists, including direct harvesting, habitat alteration, and introduction of exotic competitors. Some of these species have become reestablished in Vermont following the return of forestland over the last hundred years, either by their own efforts or by the efforts of humans to restore natural conditions. Other species remain absent from the Vermont landscape. Yet forests are not the only natural communities that occur in Vermont. In many locations, determined by soil, climate, topography, or disturbance, treeless or open communities have developed and are the communities to which we now turn.

8

Terrestrial Open Communities

NOT ALL TERRESTRIAL communities in Vermont are forested. The earliest terrestrial community following the end of the ice age was open tundra. Woodlands and forests largely replaced tundra in this region around 12,000 years ago, so that today, unless something prevents it, the community type established in most places is forested. However, several different conditions can lead to an inability of woody plants to become established widely; as a result, many different open communities have remained or have developed here since the retreat of the glaciers. Although under natural conditions they probably never in the last eleven millennia made up more than 5 percent of the total area of Vermont, they display a richness of biological diversity that contributes to the region's character.

Open communities are also of interest because of the dramatic changes they have experienced in the last 250 years. The clearing of forests and expanded settlement by European colonists beginning in the last quarter of the 1700s led to dramatic increases in some of these communities, altering the abundances and distributions of many species of plants and animals from what they had been a hundred years before. This transition also set into motion several changes for resident species. Understanding how these transitions fit into the long-term character of this naturally forested region —which is still being altered by other cultural forces, such as the spread of exotic species, acid rain, and global warming—remains a challenge.

By definition, terrestrial open communities are found on land where trees do not grow or are not well established. A host of environmental conditions can contribute to this, which results in a wide variety of open-community types. Conditions that can prevent trees from becoming established include flooding, scouring by ice, extreme cold, frequent frost and ice, deep and long-lasting snowpacks, high winds, frequent fire, unstable

TABLE 8.1.
Terrestrial Open Communities of Vermont

Shoreline communities
Riverside outcrops
Cobble shores
Riverside sand and gravel
River-mud shores
Erosional bluffs
Sand beaches and dunes
Shore grasslands
Upland meadow communities
Alpine tundra
Boreal outcrops
Temperate outcrops
Grassy meadows
Cliffs and rockfall communities
Cliff faces
Talus fields

Note: Modified from a classification system used by the Vermont Agency of Natural Resources.

ground, absence of soil, and frequent and severe human modification of the landscape. Open communities also exist as transitions between some aquatic and forest community types.

The diversity of conditions that prevent widespread establishment of woody vegetation leads naturally to a diversity of open community types, including those associated with shorelines of rivers and lakes, upland meadows and meadowlike clearings, and cliffs and rockfalls (see table 8.1). These communities can have dramatically different plants and animals associated with them and in terms of their composition can have little in common with each other, other than a general absence of trees. Yet all of the ecosystem processes that operate in forest communities (see chapter 7) also operate in these communities, controlling their rates of growth, responses to disturbance, and patterns of interactions among the species that live there.

Shoreline Communities

Along the shores of lakes and rivers, away from areas with permanent or frequent standing water, several different shoreline communities can develop. These communities are kept open mainly by flooding, ice scouring, and wind, all of which prevent woody plants from becoming well established. The identities, characteristics, and amount of vegetation in these communities are determined by the characteristics of whatever soil is able to develop there, especially the size and kind of the soil particles—ranging from clay to silt to sand to gravel—as well as by the soil's chemistry and moisture content. Shoreline communities often grade into one another perpendicular to the waterline, so that several can be found right next to each other in a very small area. Conditions that lead to their formation are common along water bodies, so these communities are widespread throughout the state. The individual sites are rarely large, though, so they make up a very small amount of total area.

Riverside outcrop communities form along the banks of rivers where flooding deposits enough fine-grained sediments in cracks in the rocks for soil to develop. All examples in Vermont are small, but they are found along all of the larger and most of the smaller rivers, even if only in a few locations. Vegetation is scant and is dominated largely by forbs and grasses. Common native plant species often create quite beautiful flowers in the spring and summer; these include harebells, Canada anemone, fringed loosestrife, joe-pye weed, goldenrod, and wild columbine. Few animals live specifically in this or other shoreline communities. The few that do are all small, ground-dwelling insects, such as the red spotted ground beetle, which forages and burrows among the cracks of these outcrops.

Where the substrate along a lake or river shore is made up of gravel and larger rocks, cobble-shore communities develop. Silt and sand are deposited between rocks by flood waters, but because the rocks are unstable, their frequent movement leads to a much different set of plant species than in riverside outcrops. These include a variety of sedges and willows, touch-me-not, and monkey flower, plus an occasional cottonwood or sycamore. The rare cobblestone tiger beetle is also found here. Cobble communities are widespread along the largest rivers in the state and the shore of Lake Champlain, their locations determined by the region's glacial history, which led to the deposition of rocks and other debris in a few specific sites.

Riverside sand and gravel communities develop where sand is deposited in areas of fast water, such as islands and sandbars in rivers (see fig. 8.1). They are sparsely vegetated, being influenced heavily by regular and severe flooding. Most of the plant species in this community type are the same as those found in other shoreline communities, but many species of mosses here are unique, being favored in this flood-prone habitat by their low-growth form and ability to tolerate being underwater for long periods of time.

River-mud shores occur in places where the water moves slowly enough for fine-grained sediments to build up over time. Vegetation on mud shores is sparse and is made up of a mix of herbs that is also found widely in other shoreline communities. Although widespread on most of the larger rivers in the state, mud shores are almost always quite small.

FIGURE 8.1. A riverside gravel community. Photo by Stephen Trombulak.

Erosional bluff communities form naturally where active soil movement occurs on a large scale, particularly where rivers cut into steep banks. Large-scale slumping of soil occurs often, and plant species that become established successfully on the banks tend to grow fast and spread out with low-lying runners over wide areas. In recent times, erosional bluffs have also developed in other areas away from rivers where human activity has exposed large banks of open ground, especially in open-pit sand and gravel excavations. Some species of birds are associated particularly with erosional bluffs, especially the bank swallow, northern rough-winged swallow, and belted kingfisher, which all build their nests in burrows in these exposed slopes.

The persistence of all these communities depends to some extent on periodic flooding, which redistributes sediments and eliminates flood-intolerant plants. The widespread development of regulatory dams threatens their persistence.

Sand beaches and dunes develop where glacial transport of sediments created well-drained, sandy areas (see fig. 8.2). Flooding, scouring by ice, and wind are all important forces here, preventing the establishment of much tall vegetation. These sand-dominated communities are found in very few places in Vermont, namely, the northern part of Lake Champlain and a few lakes in the Northeastern Highlands. Because these areas are all extremely popular recreation sites, they are very heavily disturbed by people and are threatened throughout the region. Heavy alteration by people is especially troublesome because the plants' low productivity and high rates of natural disturbance make these communities very slow to regenerate. Many of the associated plant species are now relatively rare; they were all probably distributed much more widely in early postglacial times when sand dunes and beaches were more common.

Where the shoreline substrate is most stable and subject to the least amount of disturbance, grasslands can form. Denser vegetation, dominated by tall grasses and other herbs, develops in these areas. Like other shoreline communities, grasslands are found throughout the state, associated to some extent with most rivers, but nowhere are they large. Although these communities make up only a very small percentage of the total area of the state, they are vitally important for the more than twenty-five species of shorebirds—including a variety of sandpipers, yellowlegs, and plovers—that migrate through this region each spring and fall as they move between their tropical wintering grounds and Arctic breeding grounds. The mostly open, wet ground in shoreline communities provides habitat to an abundance of burrowing insects and other invertebrates, which are the preferred food of shorebirds. With their long, sensitive bills, they probe down into the mud, silt, and sand, replenishing their body fat as they continue their long journeys.

FIGURE 8.2. A sand-dune community on Lake Champlain. Photo by Stephen Trombulak.

In addition to a diverse set of over 250 native plants, shoreline communities today support a large number of exotic plants, including such well-known species as Queen Anne's lace, yarrow, purple loosestrife, Japanese knotweed, and coltsfoot. The success of exotics in these communities is largely because they tend to survive well where there is a lot of disturbance and relatively little competition with native species, conditions found easily in shoreline communities. Therefore, in terms of species composition, shoreline communities are among the most altered of the open community. Because exotics are often fierce competitors, they make it more difficult for native species to grow, reducing these already uncommon species even further and putting them at great risk of extirpation.

Upland-Meadow Communities

Away from the forces of scouring ice and floods along shorelines, other physical factors can prevent or limit the establishment of trees. Four general types of natural meadow and artificial upland meadowlike clearings are found in Vermont, but with the exception of open fields created by humans, nowhere are they common.

Perhaps the rarest of all community types in Vermont today is alpine tundra (see fig. 8.3). Tundra communities are found only on the summits of Mount Mansfield, Camel's Hump, and Mount Abraham in the northern Green Mountains, and altogether they cover less than three hundred acres. These communities form under the harshest weather conditions in the state: air temperatures that stay below freezing for many weeks at a time, deep snowpacks that are the first to accumulate in the autumn and that last until late spring, and frequent hurricane-force winds. Immediately below the alpine tundra lie the krummholz forests (see chapter 7), made of trees that are stunted and gnarled from the extreme weather of high elevations. The forty or so species of alpine-tundra vegetation grow where even the krummholz cannot.

During the earliest postglacial times, alpine-tundra vegetation was widespread throughout the area. At the edges of the retreating continental glacier, the climate was cold, the ground bare with little soil, and the land open to the wind for thousands of miles. Low-lying plants adapted to these harsh conditions were the only ones that could survive in this newly exposed landscape. Now, 12,000 years later, tundra plants are quite rare, and many are threatened with extirpation in Vermont. Of the eight plants known to have been extirpated from Vermont, six—hairy arnica, alpine milk-vetch, purple crowberry, northern toadflax, alpine smartweed, and white camas—were part of the alpine-tundra community.

FIGURE 8.3. An alpine-tundra community. Photo by Stephen Trombulak.

Alpine-tundra plants possess special adaptations to survive in the harsh conditions of extreme cold and low moisture availability. Many, like mountain sandwort and mountain cranberry, are short and clumped together to stay below the snowpack during the winter and to retain in their dense foliage what little moisture is available to them, chiefly during the summer, when the water is not frozen. Other species, such as wavy bluegrass, alpine sweetgrass, and alpine bilberry, have adaptations that help their leaves retain water, such as being leathery, hairy, or waxy.

No known animals are unique to Vermont's alpine tundra. Many species of rodent, bird, and insect can be found in the alpine zone, but all of them are thought to move into these areas from forests below. Deer mice, red-backed voles, and Bicknell's thrushes, for example, are all present in alpine tundra but are part of more widespread populations centered at lower elevations. Many species of upland birds migrate through the tundra in the spring and fall, using the ridgeline of the Green Mountains as a geographic landmark to help guide them on their north-south journeys.

On outcrops and summits of exposed bedrock in the state's cooler regions, boreal outcrop communities develop. They are small in size, usually less than an acre or two, and are scattered widely, particularly in the northern Green Mountains and the Northeastern Highlands. They are typically surrounded by spruce-fir forests, but the exposed bedrock at these sites does not retain moisture very well nor allow trees to take root easily. A variety of grasses, ferns, lichens, and mosses are found here, along with a scattering of small trees—mostly white pine, red spruce, and balsam fir—and shrubs such as honeysuckle, blueberry, and huckleberry.

The warm-climate counterpart to this community is the temperate outcrop community. As with boreal outcrops, these form anywhere that exposed bedrock and the resulting harsh growing conditions make it difficult for trees in the surrounding forests to become established or to grow to any great size. Temperate outcrops are always small but are scattered widely throughout the warmer regions and lower elevations of the state. The plants in this community are similar to those found in boreal outcrop communities but generally reflect adaptations to milder climate conditions. Species composition varies, depending on the acidity of the bedrock.

A fourth type of upland-meadow community is the grassy meadow. Under natural conditions, this community would probably not cover a large amount of area in Vermont. Whenever good soil and mild climates exist, succession promotes growth of forest communities after a very short period of time. Perhaps the best example of a natural grassy meadow is found in areas where beavers have dammed a stream and the resulting flood has killed the trees. Any subsequent lowering of the water level of the beaver pond (as will occur every five to twenty years after the beavers exhaust their food supply of accessible hardwoods and leave for new homes) will allow open ground to emerge, often made of thick, nutrient-rich soil, perfect for the rapid establishment of a diverse community of grasses, sedges, forbs, and shrubs. Grassy emergent meadows are found throughout the state. If flooding does not occur repeatedly, emergent meadows can be an initial step away from a more marshlike community (see chapter 9) toward the reestablishment of a forest. Over two hundred species of native grasses, sedges, and other herbs, as well as a host of animals, can be found in such open grassy meadows.

Of all the birds, tree swallows are most commonly seen in grassy emergent meadows. Diving and whirling in large numbers, these aerial insectivores sweep up insects throughout the day from the dense populations that develop in the warm and wet conditions of these grassy fields.

European colonization led to the expansion of this community far beyond its naturally limited distribution. As forests were cleared for pastureland, grassy fields came to cover over 60 percent of the state. Even with the reestablishment of forest cover, some parts of the state, particularly the Champlain Valley, the Valley of Vermont, and the southern Piedmont near the Connecticut River, are today still made up predominately of these fields (see fig. 8.4) and have played an important part in supporting Vermont's tourist economy over the past forty years (see chapter 6). Many fields are in

FIGURE 8.4. An old-field pasture community. Photo by Stephen Trombulak.

active agricultural production, such as corn and pasture for cows, but where farming is no longer being practiced, a diverse old-field community develops as an initial stage toward the reestablishment of forests. Although old fields are suitable habitat for many native species of grasses, herbs, and shrubs, high levels of disturbance make them attractive for exotic plants and escaped cultivars. In fact, exotic plants—such as orchard grass, common timothy, sheep sorrel, common chickweed, shepherd's purse, ribgrass, and yarrow—outnumber native species in this community type by almost 40 percent.

Open fields that are not used for row crops create ideal habitats for many animals. The open ground, free of dense vegetation, makes it easy for birds of prey to hunt for small animals. Red-tailed hawks draw lazy circles in the sky above fields or perch upright in the occasional lone pasture tree. American kestrels, the smallest of all the falcons in North America, and more rarely loggerhead shrikes perch quietly on roadside telephone wires, waiting to spot movement in the grass below that would signal the presence of a rodent or shrew. Turkey vultures, with their wings held stiffly upward, circle endlessly in search of carrion. Turkey vultures are a recent addition to the region's avifauna, being practically unknown in Vermont prior to the late 1930s. It is thought that their spread into Vermont is due to an increase in the number of animals killed on roads (one of the many adverse ecological consequences of an increase in roads and automobiles) and an increase in white-tailed deer mortality due to their overpopulation.

During winters when prey populations in the Arctic plummet, snowy owls come south into the Champlain Valley and Northeastern Highlands to feed on birds and small mammals from fence-post perches. The northern harrier is perhaps the most noticeable of all the open-field birds of prey. With wingspans of up to four feet, these large birds fly back and forth across the fields, often no more than a few feet above the ground, in search of small mammals.

Many other birds soar above these open fields. The courtship flight of the common nighthawk is readily observed, especially in the southern Piedmont. To attract the female, the male makes large diving loops from a great height; as he nears the bottom of his dive, wind passing through the feathers of his wings makes a loud boom, heard over great distances. Barn swallows, cliff swallows, and chimney swifts are also found in the skies over open fields.

Many species of bird prefer to spend more time on the ground. One of the most beautiful of all open-field birds is the eastern bluebird. This sky-blue bird with chestnut breast and long wings has recently declined throughout its range, primarily because it has been losing nesting holes to the introduced European starling. Starlings were introduced into Central Park in New York City in 1890 and 1891 by a group of people who wanted to bring to the United States all of the species of birds mentioned in the works of Shakespeare. The one hundred starlings they released inaugurated a spread to every part of North America below the Arctic, causing dramatic declines in native hole-nesting birds. Starlings were first seen in Vermont in 1913. The eastern bluebird has recovered to some extent through the widespread use of bluebird nest boxes, placed in fields and yards throughout their range by people concerned with the bluebird's welfare. Starlings are also responsible for declines in other native bird species, such as the purple martin and redheaded woodpecker.

Other native species of grassland birds include savannah sparrow, vesper sparrow, grasshopper sparrow, horned lark, eastern meadowlark, killdeer, bobolink, and mourning dove. During the winter, the bare fields support a group of birds that migrate here from further north. Large flocks of snow buntings and Lapland longspurs, in particular, are seen commonly on the bare or snowy ground.

Several open-field birds are new to Vermont. Ring-necked pheasants, native to Asia, were introduced by the thousands into the Champlain Valley in the late nineteenth century. They were considered such a potentially important game bird that the Vermont Fish and Game Service ran a game farm for raising pheasants during the late 1920s and early 1930s. Pheasants, however, do not easily tolerate Vermont's occasional severe winters, even in the milder climate of the Champlain Valley, and a large self-sustaining population of pheasants has never become established. Although they are seen occasionally along the edges of roads in agricultural fields, their numbers are quite low, and they are probably maintained only by constant replenishment from captive stocks.

Rock doves, more often called pigeons, were introduced into North America by French colonists in the 1600s. House sparrows were introduced in several places in North America throughout the mid- to late 1800s, including Vermont between 1874 and 1876. Originally it was thought that house sparrows would control the rising populations of many agricultural insect pests, but in fact house sparrows prefer to eat seeds. The spread of agriculture and livestock during this time gave the growing house-sparrow populations plenty of food, and they spread quickly throughout the continent.

Perhaps the most interesting of all new arrivals is the cattle egret. A small population was apparently blown by a storm from their native Africa to southeastern South America in the late nineteenth century, where they found an abundance of open fields and plains, their preferred habitat. Since then, the cattle egret has expanded its range northward through northern South America, Central America, and North America, first arriving in Vermont in 1961. This long-legged, all-white bird is seen occasionally in small flocks throughout the spring and summer in fields in the Champlain Valley, feeding on insects fleeing tractors or foraging cows. Many other birds associated more commonly with aquatic communities forage in open fields, especially the Canada goose, snow goose, ring-billed gull, and herring gull. Some species that are found more commonly in forests, such as wild turkey, American woodcock, flycatchers, and various warblers, emerge into fields along the forest margins or along the hedgerows that divide farm fields.

Because the dominant natural communities of this region over the last 10,000 years have been forested, it is questionable how abundant any of these open-field species were before widespread modification of the landscape by European colonists. Quite likely, the open-field birds that now populate much of Vermont's landscape are only recently added elements of Vermont's fauna.

A few species of small mammals can be found in open fields, although none of them is restricted exclusively to this habitat. The meadow vole, deer mouse, meadow jumping mouse, and woodchuck are among the few native species that can be found there. Also present are two species of rodents brought to North America from Europe by early settlers: the Norway rat and house mouse. Although these species do not often live far from buildings, they can be among the most common mammals around farms. Also introduced to Vermont is the Eastern cottontail rabbit, native to southeastern North America. It is not clear when it first came into Vermont, but through the twentieth century it has grown in numbers to the point where it might well have completely eliminated the native rabbit, the New England cottontail.

Another newcomer to Vermont is the coyote. This small doglike animal is native to open plains in central and western North America. Yet with the spread of agriculture and elimination of its larger competitor, the timber wolf, the coyote has steadily expanded its range throughout the East, becoming numerous in Vermont over the last fifty years. Although they are primarily active at night, they can occasionally be seen trotting in their characteristic "tail down" posture through open fields.

The relationship among coyotes, wolves, and domestic dogs is confusing because all of these species interbreed so easily with each other. Wild coyotes are known to contain wolf genes, and wolves, even those from remote regions of Canada, contain coyote genes. Both of these species look and behave quite different from each other: Wolves are large, reaching one hundred pounds and live in packs of ten or more, whereas coyotes reach only about fifty pounds and live in smaller groups of three or four. But since postglacial times, cross-breeding has occurred, and so too with dogs. Domestic dogs are distinct from their wild relatives in appearance and behavior but can interbreed with them. Such hybrids, usually seen only in captivity, pose challenges to humans. Hybrids often display a great degree of tameness because of the genes of their domesticated ancestors, but they also have the potential to be quite ferocious, leading to unfortunate consequences and the development of an unnecessary fear of wild canines among humans.

White-tailed deer benefit from open fields. Although they are primarily animals of woodlands, their preferred foods are the shrubs that grow up at the edges of field and forest, where the increased sunlight allows forest understory plants to grow rapidly.

A couple of snake species in Vermont are found regularly in open fields. The milk snake, with its distinctive inverted-Y on the back of its neck, is common in the lowlands of the Champlain Valley and the Connecticut River Valley. The smooth green snake is perhaps the most beautiful of all the snakes in Vermont; its body is almost entirely an iridescent green. Its numbers appear to be declining in recent years for reasons that are not known. One speculation is that the increased use of machines to harvest hay makes it more difficult for snakes to escape being cut or caught up in hay bales.

Few amphibians are found in open fields because their thin skin makes them prone to drying out when exposed to the direct sun. Two exceptions to this are the American toad and Fowler's toad. The American toad is familiar from fields and lawns throughout Vermont. Its gray-green, warty skin is thicker than that of other amphibians, making this small hopping animal better able to live away from water. The much rarer Fowler's toad is known to live in only a few locations in the Connecticut River Valley of southern Vermont. It looks much like the American toad but has more and smaller warts and an unmarked white belly.

Of all the insects of open fields, butterflies are the most noticeable, as they fly from flower to flower in search of nectar. The diversity of colors to be seen is suggested by the names of some of the more common species: silver-bordered fritillary, mountain silverspot, pearl crescent, Baltimore checkerspot, Milbert's tortoise shell, orange sulfur, red admiral, painted lady, bronze copper, and black swallowtail.

Perhaps the most interesting of all the butterflies here is the monarch. This large orange and black butterfly makes a unique long-distance migration that begins in the cool nights of autumn. Up to one hundred million monarchs east of the Rocky Mountains fly south and spend the winter densely aggregated in small patches of fir forests in the mountain of southern Mexico. The following spring, these butterflies fly as far north as the southern United States, where they lay eggs in fields on milkweed plants before they die. The butterflies that eventually hatch from these eggs continue the journey northward, on which up to five generations may live and die before the winter migration to the south begins again. How these animals know where to go in their travels remains a mystery.

Grassy fields are also home to a rich diversity of other insects, including beetles, crickets, locusts, aphids, leafhoppers, walking sticks, and mantids. Many of these are colored cryptically to blend in with the vegetation and avoid being eaten by birds. Others use colors to mimic less-palatable species. For example, robber flies and drone flies both have yellow stripes that make them look like bees, and soldier flies look very much like yellowjackets.

Cliff and Rockfall Communities

With their vertical slopes, cliffs and ledges are home to a sparse and unique set of species adapted to the chemistry and climate of these severe habitats (see fig. 8.5). Many of the state's rarest species are found here, perched in cracks where small amounts of soil accumulate. A few trees, shrubs, and herbs characteristic of the surrounding forest can take root here, but these cliff and ledge faces also support about fifteen species of plants that are found no other place, such as bulbet fern, fragrant fern, smooth woodsia, purple cliff-brake, rock polypody, and walking fern. Variations in climate and the chemistry of the rocks greatly influence exactly which plant species make up these communities.

FIGURE 8.5. A cliff community. Photo by Stephen Trombulak.

A few bird species are associated especially with cliffs. Ravens, the largest of the crowlike birds in North America, prefer to make their large stick nests on cliffs, where inaccessibility protects their eggs from predators. They can often be seen flying and playing in small flocks, diving and soaring while calling to each other in low croaks. Ravens were, at one point, extirpated from virtually all of New England. The dramatic declines in undisturbed forest habitat as well as the extirpation of deer and moose (the carcasses of which make up a large part of the raven's diet) undoubtedly contributed to this decline. Only one report of a raven was made in all of Vermont between 1912 and 1961. With the recent increases in forestland, deer, and moose, ravens have made a remarkable recovery. Beginning in 1961, their populations have grown steadily, and they are now year-round residents throughout the state.

Cliffs are also home to peregrine falcons, native birds of prey that were once extirpated from the state and drastically reduced everywhere throughout their range. Their decline, which might eventually have resulted in their complete extinction, was due largely to reproductive failures caused by the widespread human use of pesticides such as DDT. These pesticides, while successful over short periods in controlling insect pests, are passed up the food web from plant to insect to insect-eating bird to falcon. DDT inhibited the ability of a female falcon to lay eggs with shells strong enough to prevent breaking during incubation, and as a result the population could not replace itself.

As a result of a ban on such pesticides, the environment is once again able to support a self-sustaining population of peregrine falcons. Peregrines began to be reintroduced to Vermont in 1977 from captive breeding populations maintained in New York. Like ravens, peregrine falcons nest on exposed cliffs. Reintroduction involves the construction of artificial nests on these cliffs, where eggs taken from nests of captive birds are hatched and the young reared by humans. Once the chicks grow up, they are returned to the cliffs to nest, thereby increasing the resident population. Today there are at least eighteen territorial pairs of peregrine falcons on cliffs in the state, such as at Deer Leap in Bristol and Bald Mountain in West Haven; most of them have successfully produced chicks for several years.

The fractured and broken rock at the base of cliffs can be too big for soil to form and too unstable for large vegetation to become established. Here, rather than talus woodlands (see chapter 7), are talus fields predominantly made up of nonwoody plants. Plants with low-lying growth forms are the most common, such as rock polypody, marginal woodfern, Virginia creeper, and a wide variety of lichens.

Talus slopes are home to only a few animals. Timber rattlesnakes were once found widely throughout the state in this habitat, but through hunting they have now been reduced to only two sites in the southern Champlain Valley (see chapter 7). Even more rare is the five-lined skink, the only species of lizard found in Vermont. These lizards reach the northernmost limit of their range in the Champlain Valley. A talus slope near Lake Champlain in West Haven is their only known site in Vermont.

Little is known for sure about the habitat and behavior of the long-tailed shrew. The few signs of this tiny, secretive animal in Vermont suggest that it prefers to live deep in the cracks and crevices formed by the talus. Because they are so small and because it is impossible to observe what is taking place deep under the boulders, these shrews have been able to avoid close study. It is likely that they live much as do the other shrews in Vermont, by feeding almost constantly on small insects, spiders, and centipedes.

As with forest communities, open communities are diverse and comprise a wide variety of interconnected species. The present-day distribution of terrestrial open communities differs dramatically from that of two hundred years ago. Some communities, such as shore beaches and dunes, have declined in number and quality over that time, whereas old fields have increased dramatically. The great expansion of agriculture in the nineteenth century resulted in increases in a host of species adapted to open-field environments, all of which are still present today and are themselves undergoing transitions as the landscape once again becomes increasingly forested. Forest and open communities have provided the context for much of the natural character of Vermont, as well as for the development of human cultures here for the past 12,000 years. But they are connected in many vital ways to places throughout the state whose characteristics are influenced less by land and more by water—the wetland and aquatic communities.

9

Wetland and Aquatic Communities

ALTHOUGH MOST OF the area of Vermont is solid, dry ground, many community types are dominated through all or part of the year by standing or moving water. These wetland and aquatic communities are important parts of the natural history of this region; they are homes to a large number of plant and animal species, they provide food to countless other terrestrial species, and they play critical roles in many ecosystem processes that take place within the region.

They have also played important roles in the history of the development of human cultures in Vermont. The use of lakes and rivers as corridors for transportation, sources of power, and sources of food was central to the establishment of human settlements and industry beginning with the earliest colonization by Paleoindians. Human settlements in Vermont have also had, and continue to have, major impacts on all wetland and aquatic communities, particularly through changes in the flows of rivers and streams, pollution, and draining of wetlands.

Because humans are terrestrial animals, however, they have a much greater general awareness of terrestrial communities. Knowledge of variations among wetland and aquatic communities and how the associated organisms are distributed is much less complete than it is for forested and open communities in Vermont. Classification of these communities is difficult in part because slight changes in the chemistry of water, duration of standing water, climate, and surrounding communities can have tremendous effects on the dominant species present. Many of these communities are best described as transitional stages between fully aquatic and fully terrestrial communities. As with shoreline communities (see chapter 8), some wetland communities intergrade with each other over short distances, making them difficult to tell apart. Yet broad differences among aquatic communities are

TABLE 9.1.
Wetland and Aquatic Communities of Vermont

Peat lands
Bogs
Fens
Marshes
Sedge meadow
Marshes
Swamps
Shrub swamps
Forested swamps
Seasonal wetlands
Woodland vernal pools
Open water
Free-flowing waterways
Ponds and lakes
Lake Champlain

Note: Modified in part from a classification system used by the Vermont Agency of Natural Resources.

obvious and form the basis for a simple classification system based on characteristics of the water and dominant vegetation (see table 9.1).

Peat Lands

In areas where water can accumulate without having an easy means to drain out again, dense mats of sphagnum moss, the primary plant that forms peat, can develop. Peat lands tend to develop where groundwater flow into an area is limited or nonexistent and where the majority of nutrients come from rainwater. In these areas, the nutrients needed for plant growth, especially nitrogen, are in short supply. Oxygen is also limited, making decomposition of organic matter slow and incomplete. In fact, this slow decomposition is what makes peat lands around the world such an excellent source of fuel; large mats of poorly decomposed organic matter build up year after year, providing a source of fuel that can be easily harvested and burned. Peat is rarely used for fuel in Vermont, however, since wood is so abundant.

Two basic types of peat land are found in Vermont: bogs and fens. In general, bogs differ from fens in that bogs derive virtually all of their water from precipitation rather than groundwater or surface flow, have very low nutrient availability, and are highly acidic. Bogs are found throughout the state but primarily in the Northeastern Highlands, associated with kettle-holes, old lakes, and other depressions, many of which are remnants of the carving of the landscape that occurred during the most recent ice age.

Vegetation in bogs is dominated by sphagnum moss, which forms a continuous cover over the water on which other plants can grow. Sometimes these mats can be massive, in some locations ranging up to fifteen feet in thickness above the bog's acidic waters, easily supporting the weight of woody vegetation and large mammals. Various shrubs can be found growing on the mats, especially laurels, Labrador tea, cranberry, leatherleaf, chokeberry, and rhodora. Bog herbs, which tend to do well where nutrients are poor, include sundew, pitcher plant, sedges, and a wide variety of colorful orchids.

Sundews and pitcher plants have unique strategies for acquiring nitrogen. Since this essential nutrient is lacking in the water that underlies a bog, these plants capture insects to gain nitrogen. Sundews are covered with sticky hairs that trap insects, while a large modified leaf of the pitcher plant forms a basin of water that traps insects who fall into it. Other insects that live in the pitcher water capture and eat the pitcher plant's prey, and the pitcher plant absorbs the nitrogen that is released by the fecal matter of the associated insects.

Some bogs have a few scattered trees associated with them, particularly black spruce, gray birch, red maple, and, at Maquam Bog in the Champlain Valley, pitch pine, which might have arisen there as the result of purposeful burning by Abenaki Indians to encourage the growth of blueberries. Bogs may be surrounded quite closely by other wetland types, such as fens, where the local topography allows groundwater to bring in more nutrients and oxygen.

Like bogs, fens are dominated heavily by mosses. Variation among fens arises from differences in nutrient availability in the water, which results in fens ranging from those comprising mostly peat moss and brown moss, where nutrients are limited, to those that are almost all brown moss and sedges. Occasional trees develop on hummocks in the peat, particularly black spruce, red spruce, tamarack, and red maple.

Few vertebrates are specific to bogs and fens, although many forest dwellers are attracted to bogs because of the ecological richness that is present along the edges of forests and bogs. Forest birds that feed on flying insects, such as the eastern kingbird, yellow-rumped warbler, and yellow-bellied flycatcher, use trees at the edge as perches to scan for prey out across the peat land's open space. The increased availability of sunlight at edges also allows for faster growth of forest shrubs and herbs, which in turn attracts many small mammals, such as meadow voles.

The southern bog lemming is probably common in bogs throughout this region. This small, shy rodent lives in colonies that form nests and complex tunnel systems in the sphagnum, preferring to feed on the leaves, stems, and seeds of the grasses and sedges that grow on the mats. By eating and tunneling even at great depths in the sphagnum, they probably play an important part in the cycling of nutrients throughout the upper portion of the peat lands.

The most numerous of all the bog species are the insects, especially butterflies. Some butterflies—such as the bog copper, bog purple fritillary, and bog elfin—specialize on plants found primarily in peat lands. A host of other invertebrate groups, including spiders, ants, centipedes, and slugs, can be found in peat lands as well, depending on how well they tolerate the high acidity of bogs water.

Emergent Wetlands

Emergent wetlands develop where water flows more steadily than it does in peat lands, yet little to no woody vegetation is present. They may be seasonally or permanently flooded, but the presence of standing water at least part of the year is enough to alter significantly the composition of the soil compared to that of surrounding terrestrial communities. Soils in emergent wetlands tend to be mucky and permanently saturated with water, thus reducing the diversity of plants that can grow there. Emergent wetlands are widespread but scattered throughout Vermont, and small patches form easily wherever water can permanently saturate the soil. For example, small marshes form quickly in urban areas and roadside ditches whenever changes in topography or diversions of storm water convert areas from being well drained to being semipermanently or permanently wet.

Natural emergent wetlands can be extremely large, especially along the shores of Lake Champlain, where water only a few inches deep in mid- to late summer spreads out over expanses of many hundreds of acres. Some emergent wetlands intergrade smoothly with communities that have more terrestrial characters, such as grassy emergent meadows. It is perhaps most useful to view grassy emergent meadows and emergent wetlands as the intermediate or transitional stages between fully terrestrial and fully aquatic communities.

Sedge meadows form where the soils are permanently saturated but the ground does not necessarily remain continuously flooded. The soil is generally mucky, composed of well-decomposed organic matter, often with some small amount of peat. Sedge meadows are somewhat transitional to more nutrient-rich fens. The dominant plants in sedge meadows are, as the name implies, sedges, but a host of other plants are present, many of which are also common in bogs and fens.

FIGURE 9.1. A cattail marsh. Photo by Stephen Trombulak.

Where standing water occurs for longer periods of time during the year, marshes dominated by cattails and bulrushes form (see fig. 9.1). Fields of marsh plants grow in dense stands along slow-moving waterways throughout the state. The high productivity of these waters make them important habitats for many species of animals. A wide range of aquatic birds are found almost exclusively in marshes, including the pied-billed grebe, American bittern, Virginia rail, sora, common moorhen, and American coot. More than fifty species of ducks and shorebirds that breed in Vermont or that migrate through it in the spring and fall depend on the aquatic vegetation and insects that marshes provide. Some marsh birds, such as the marsh wren and red-winged blackbird, are not themselves aquatic but make their nests in the cattails and rushes where it is difficult for predators to get to them, and they feed their young from the marsh's abundant supply of insects.

Red-winged blackbirds are perhaps the most noticeable of all the marsh birds. Like many other species of birds, male and female red-winged blackbirds look noticeably different. The females are colored with brown and white streaks, making them difficult to see amid the emergent marsh vegetation as they move back and forth from their nests to the water in search of food. The males, however, are distinctly colored, being jet black over all of their bodies except for bright-red shoulder patches. The males defend territories, in which the females build their nests, by clinging to upright stems at the territory's edge and flashing their shoulder patches at neighboring males, while emitting a loud call of "conk-ka-ree." Males establish territories almost anywhere marsh vegetation develops, including in small clumps of cattails in drainage ditches along the sides of roads. Although red-winged blackbirds have traditionally been migratory birds, heading south for the winter and not returning until food is again available in the spring, in recent years varying numbers have remained in the region year-round depending on the amount of food available near farms during the winter.

The majestic great blue heron, standing up to four feet tall, is often seen wading slowly through marshes in search of fish. With its long neck and spearlike beak, the heron darts its head quickly into the water to grab small- to medium-sized fish that swim among the marsh's emergent vegetation. Great blue herons are also associated closely with forest communities; they build large stick nests in tree tops in dense colonies in many places throughout the state, especially northern hardwood forests in the Green Mountains, on islands in Lake Champlain, and in floodplain forests along the shore of Lake Champlain.

The muskrat, a close relative of the meadow vole and bog lemming, is a prominent marsh dweller. This large semiaquatic rodent, which can reach more than two feet in length, has a tail that is slightly flattened from side to side, which it uses to propel itself forward through the water in search of young succulent plants and, on occasion, aquatic animals such as insects and fish. Muskrats make their dens in banks along the sides of marshes as well as directly within the marsh itself in domelike structures of cattail stalks.

Marshes are also home to a number of aquatic reptiles and amphibians. Snapping turtles, among the largest and most aggressive of all the turtles in North America, swim or wait quietly for fish, crayfish, or any other prey to come by. Water snakes are the most aquatic of all the snakes in Vermont. Although in Vermont they can be found in other aquatic-community types, they are known primarily from marshes and ponds in the Champlain Valley. Like snapping turtles, they spend their time searching for small animals among the emergent vegetation and along the bottom sediment.

Emergent wetlands are linked intimately with many ecosystem processes critical to the overall health of the landscape. Because of the presence of dense stands of vegetation, water that flows through wetlands tends to move very slowly. This gives any suspended sediments an opportunity to settle out. The deposition of sediment in the marsh then helps to build up the land, as well as to prevent water that flows through the marsh from carrying large amounts of eroded soil out of the area and, perhaps, into a lake where they could damage the plants and animals there.

Wetland soils and plants can absorb a great deal of water during times of high water flow, such as after rainstorms and in the spring when snow is melting. The ability to absorb water quickly and then release it slowly over time, much like a sponge, is a critical aspect of natural flood control in all of the low-elevation regions of the state. Also, the native plants that grow in emergent wetlands are particularly efficient at absorbing pollutants and excess fertilizers that enter surface water from urban and agricultural activities in the upland areas surrounding the main channel of the water flow. The ability of these plants to purify water has important implications for the plants and animals that live downstream of these wetlands, both those that live directly in the water and those that live on the land but depend on wetland and aquatic communities for food.

Unfortunately, emergent wetlands have been dramatically reduced throughout the region, especially in areas with soils that have great agricultural potential. By the 1980s, 35 percent of the wetlands present during presettlement times had been lost. Although the incremental drainage and filling of wetlands that has gone on here over the last two hundred years might at first seem beneficial because it helps to create land that can be put to "productive use," it actually increases the chances of flooding, erosion, siltation, and pollution of downstream waters. Most of the remaining emergent wetlands have been invaded recently by exotic species—especially purple loosestrife, flowering rush, and yellow iris—that are choking out many native species.

Swamps

Where land is permanently flooded with shallow water or has a high water table but the soil is well developed enough to support woody vegetation, forested wetland communities called swamps develop (see fig. 9.2). These vary based on the height of the woody vegetation, from shrub swamps to forested swamps, and on the types of tree species present, from hardwood-dominated to softwood-dominated swamps. Swamps are found widely throughout low-relief areas in the region, especially associated with rivers and river-bottom valleys in the Northeastern Highlands and Champlain Valley. They can be large or small, depending on how the local topography controls the flow of water. Some swamps spread over very large areas. Cornwall Swamp, dominated by red maple and northern white cedar, in the Champlain Valley near the Otter Creek, covers more than fifteen square miles.

Which species dominate in a swamp depends largely on the chemistry of the soil and climate. Swamp soils are all, as a rule, completely saturated, but they have varying amounts of organic matter and nutrients. Where soils are low in nutrients, plants more typical of bogs are common. The presence of trees in these communities leads to the development of many microsites

FIGURE 9.2. A forested swamp. Photo by Stephen Trombulak.

that are well above the water. Root mounds of living trees and decomposing trunks of fallen trees form raised hummocks that provide sites for many plants that prefer drier conditions.

A host of dominant trees are found in various swamps, including red maple, black ash, black spruce, tamarack, black gum, white cedar, green ash, swamp white oak, silver maple, speckled and smooth alder, buttonbush, a variety of willows and dogwoods, and an occasional elm. Where the climate is colder, coniferous species dominate; where warmer, swamps contain primarily hardwoods. The understory and herb layer can be sparse, but many swamps are rich with almost two hundred different species of small trees, shrubs, and herbs, particularly sedges, grasses, mosses, and ferns.

Many of the animals that live in marshes are also present in swamps. The presence of water and emergent vegetation provides a similar habitat and, therefore, supports similar fauna. Spruce swamps in a few locations in the Northeastern Highlands have small populations of birds that are found nowhere else in the state, in particular the spruce grouse, three-toed woodpecker, black-backed woodpecker, and gray jay. Why these species are restricted to coniferous forests in the Northeastern Highlands and do not extend into similar forests elsewhere in Vermont is not well understood.

As with emergent wetlands, swamps play an important role in controlling erosion and flooding, as well as in removing pollutants from water as it flows downstream. They have been reduced over much of the state since the time of European colonization. Their predominant locations near waterways in areas of low relief make them attractive sites for buildings and agricultural fields.

Seasonal Wetlands

Among the least appreciated but ecologically most important wetlands are those that have standing water for only two or three months in the spring and early summer, enough time so that small aquatic communities can develop but not enough so that the soil becomes permanently saturated. The best known type of seasonal wetland in Vermont is the vernal woodland pool. In shallow depressions in a variety of terrestrial communities throughout the state, snowmelt and spring rains collect to form small, temporary ponds. Because they do not last throughout the year, fully aquatic predators, mainly fish, cannot live there, and vernal pools become important breeding sites for many species of frogs, toads, and salamanders.

The best known of all the vernal pool breeders in this region is the spring peeper. The males of this tiny, inconspicuous species of frog live on the edges or in the emergent vegetation of vernal pools, as well as marshes, swamps, and ponds. Individually, each male makes a high-pitched whistle-like peep with its inflated vocal sacs, repeated every five to ten seconds. The call of spring peepers is one of the first signs in Vermont that spring has arrived. A large group of males all calling at the same time, known as a chorus, can be heard over very long distances and sounds like a continuous whistle; such a chorus might last for hours, from dusk until late into the night. Females are attracted to the breeding grounds by calling males. Each female will select one male to mate with, deposit several hundred eggs into the water and, as the male fertilizes them, attach them to submerged vegetation.

After breeding, both male and female peepers disperse into the surrounding forests to live inconspicuously in the leaf litter and vegetation of the forest floor, feeding on small invertebrates. The eggs of the spring peeper hatch in six to twelve days, and the tadpoles go through their entire development and metamorphosis in about three months so as to be ready to live on land before the vernal pool disappears in the warmer temperatures of summer.

Many species of salamander breed in vernal pools, including two closely related species of mole salamanders, the Jefferson salamander and blue-spotted salamander, which in the past produced unique intermediate hybrid forms. After breeding, mole salamanders, like spring peepers, disperse away from the vernal pool; but rather than live in the leaf litter and vegetation, they burrow deep underground to feed on subterranean insects and other invertebrates.

Many other amphibians come to vernal pools to breed. Some, like the American toad, spotted salamander, and wood frog, are found widely throughout the state, in all regions and at all forested elevations. Others are more restricted in their distribution: The four-toed salamander is present only in the Champlain Valley, Valley of Vermont, and southern Piedmont at elevations less than 1,200 feet. The western chorus frog is known from only a few records of choruses at the northern end of the Champlain Valley.

Free-flowing Waterways

Peat lands, swamps, and emergent and seasonal wetlands all have vegetation that sits on or rises up out of the water and have water flows that are generally slow or nonexistent. Aquatic communities develop where emergent vegetation, because of depth, water velocity, or substrate, generally does not grow. In some of these areas, water flows more rapidly in distinct channels. There are more than five thousand miles of free-flowing water in Vermont. Rivers and other waterways, more than any other single factor, determined the early pattern of human migration and settlement in Vermont (see chapter 3).

The biological characteristics of free-flowing waterways are influenced strongly by characteristics of their physical structure. One group of waterways in Vermont is found where steep slopes force water to flow at high speeds, and therefore leave a bottom substrate that is dominated by large boulders, cobbles, and gravels (see fig. 9.3). The primary distinguishing biological characteristics of these high-gradient waterways are the insects and fish that live there. They tend to be species that can cling easily or hover close to the bottom of the channel so as not to get swept downstream; they also prefer to live at the generally colder temperatures of these waters. The aquatic insects in high-gradient waterways are almost all the larval stages of the species; most metamorphose into terrestrial, usually aerial, adults. High-gradient waterway species include small winter stoneflies, green stoneflies, caddis flies, brook trout, slimy sculpin, and blacknose dace.

A few species of salamanders are specific to streams and rivers. The spring salamander, uncommon but widespread throughout the Green Mountains, is bright red, reaching up to seven inches in length. It forages for small aquatic animals from the banks, on rocks in the water channel, or even in the water itself. Smaller and more cryptic are the northern two-lined salamander and dusky salamander, both of which prefer to live alongside or in high-gradient streams.

FIGURE 9.3. A high-gradient waterway. Photo by Stephen Trombulak.

High-gradient waterways vary in width and water volume but are almost always associated with mountains. Examples of these include virtually all of the headwater streams and brooks that flow off of the Green Mountains and the Taconic Mountains, as well as many of the rivers that flow into the Connecticut River, such as the Passumpsic, Nulhegan, and White Rivers.

In places where slopes are shallower, particularly in the lowlands of the Champlain Valley and the Valley of Vermont, the flow of water slows down and more fine-grained sediments, such as clay, silt, and sand, are deposited on the bottoms (see fig. 9.4). Low-gradient streams and rivers are often rich with aquatic species, especially those that tolerate warmer temperatures and larger amounts of sediment. More than ninety species of freshwater fish live in Vermont, and almost all of them spend at least part of their lives in low-gradient waterways. Species within some fish families, particularly minnows, catfishes, sunfishes, and perches, segregate out into waters with slightly different conditions. For example, blacknose dace and common shiner are generally found where cool high-elevation waters mix with warm waters at low elevations. The closely related bluntnose minnow and creek chub, on the other hand, prefer more uniformly warm waters. The exact distribution of the freshwater fish in Vermont and their habitat preferences are still not completely known.

Members of one group of fish in the Champlain Valley, the lamprey, spend part of their lives in Lake Champlain and part in the lower reaches of

FIGURE 9.4. A low-gradient waterway. Photo by Stephen Trombulak.

some low-gradient waterways that flow into the lake. The best studied of these is the sea lamprey. When adults, these jawless fish live in Lake Champlain, where they feed as ectoparasites on other fish. After living in Lake Champlain for two or more years, adult lamprey migrate up certain waterways, such as the Poultney River and Lewis Creek. Where the substrate is made of gravel, a male creates a small depression in the stream bottom, into which a female deposits more than 200,000 eggs. The adults then die. After the juveniles, called ammocete larvae, hatch, they migrate downstream to where the stream bottom is made of finer sediments, where they bury themselves tail first until just their heads stick up into the water column. Ammocetes live as sedentary filter feeders for five or more years until they eventually metamorphose into adults and migrate downstream to enter Lake Champlain.

Adult sea lamprey feed by grasping onto the sides of such thin-scaled fish as lake trout. Lamprey use their tongues to rasp wounds in the sides of their hosts and to feed on blood and body fluids. After a time, the lamprey drops off; although one or a few lamprey attacks are not lethal, if the density of lamprey is great enough, then the frequency of attacks can be enough to weaken and eventually kill the trout. Concern over the impact of lamprey on large fish in Lake Champlain, especially those like lake trout that have economic value as game fish, led to a U.S. Fish and Wildlife Service program in the late 1980s in which lampricides, chemicals that kill the larval lamprey, were put into waterways near the sites of the larval feeding areas. This resulted in a dramatic decline in sea lamprey and an increase in their host species in Lake Champlain. The program was not without controversy, however, as the lampricides also killed a variety of other aquatic species. The ethics of the lamprey-control program is complicated by a lack of certainty over whether sea lamprey are native or exotic to Lake Champlain. One theory has it that the sea lamprey entered into the lake via the Champlain Canal after its completion in 1822. Another theory holds that lamprey are native, having entered the lake during the time of the Champlain Sea or from the Saint Lawrence River via the Richelieu River. Natural-history accounts prior to the early 1800s are incomplete, and perhaps the true status of the sea lamprey in Vermont will never be known.

Low-gradient waterways in Vermont support several groups of invertebrates, including insects, snails, crayfish, and mussels. Seventeen species of native freshwater mussels are found in Vermont, virtually all of them in low-gradient rivers and streams. They are all sedentary filter feeders, using large feeding apparatuses to collect small plants and animals out of the water. They, in turn, are food for many larger species, particularly mammals that forage in and around these waterways.

A few species of birds and mammals are associated with streams and rivers. The belted kingfisher perches on vegetation on riverbanks, looking for fish in the water below. When they spot a fish, they dive quickly with their long, sharp beaks pointing straight down to grab the fish.

Many predominantly terrestrial mammals live along the edge of these waters. The water shrew, mink, and raccoon all hunt for prey among the vegetation at the water's edge. The river otter, once numerous in many rivers, especially in the Champlain Valley, has been reduced to very low numbers by overtrapping and other disturbances. These semiaquatic mammals hunt for fish and invertebrates in the water, dipping and diving up and down the shores. Their thick pelts, which protect them from the cold by trapping insulating air in the hairs, made them prime targets for fur trappers. Although they have not been extirpated, their numbers have been greatly reduced, and they are not often seen.

Another semiaquatic mammal whose pelt was prized by trappers is the beaver. This is the largest of all the rodents in North America, sometimes exceeding sixty pounds. Next to humans, they probably alter their environment more than any other species of mammal in North America. Using their large front teeth, they cut and transport small trees and branches to build large dams in low-gradient streams. Water builds up into ponds behind these dams, which provide beaver protection from predators and easy access to many more trees and shrubs. During the summer, beaver prefer to eat aquatic plants such as water lilies and cattails. In the winter, however, when these are not available, they eat the cambium under the bark of trees, preferably aspens and cottonwoods, but also a range of other hardwoods and, as last resorts, softwoods, and will spend a great deal of time cutting and transporting woody vegetation to their dens to feed upon later. Although beaver also live in banks along large rivers where dams are impossible to make, they are most noticeable in beaver ponds, where they build lodges made of piles of sticks up to six feet in height. These dens are surrounded by water, usually with just a single underwater entrance.

Beaver were once abundant throughout North America. During the 1600s and 1700s, however, their pelts were highly prized for the fashion industry in Europe, making beaver one of Vermont's first links to international commercial economies, and they were trapped almost to extinction. They were completely extirpated from Vermont by 1850. Beaver were reintroduced in the southern part of the state in 1921, and with the return of the forests to provide them with food, their populations have grown steadily to the point where they are once again found throughout the state. Unfortunately, the forest communities that the beaver were reintroduced into were not the same as the ones they had lived in before. With the extirpation in the 1880s of the timber wolf, the beaver's main predator and the primary check on its population was gone. Beaver populations have probably increased far beyond their historical levels and have saturated available habitat, and many people are concerned that without the reestablishment of a natural predator-prey relationship, beaver will leave few low-gradient streams in Vermont unmodified.

The wood turtle also lives in or near woodland rivers and streams. This quiet, slow-moving reptile has a heavily sculpted shell and is found uncommonly but widely throughout Vermont. It spends the winter hibernating in the mud of stream banks and bottoms. During the summer it roams widely in search of a mixed diet of vegetation, fungi, insects, snails, and carrion. Populations of wood turtles have declined in recent years, many of them hit and killed by cars as they slowly wander across the roads that now crisscross the state.

Of all the free-flowing waterways in Vermont, the Connecticut River is in a class of its own. Wide and deep, it flows through the Northeastern Highlands from its headwaters in New Hampshire and through the Piedmont on its way to the Atlantic Ocean in Long Island Sound. The Connecticut River supports a mostly warm-water fish community, but at one time it included major spawning runs of Atlantic salmon and American shad.

Rivers and streams have been altered significantly by human settlements. Many towns and villages take their drinking water directly from these surface waters, and many streams that historically had continuous flows today dry up entirely during the summer months. Ski resorts in the Green Mountains have also come to rely heavily on water diverted from these natural flows to make artificial snow in order to extend the ski season. Rivers and streams have been dammed in many places (see map 5.1) to generate electricity and to control flooding, altering flow regimes and harming many aquatic animals that depend on free-flowing water for their existence. For example, beginning in the 1930s, hydroelectric power dams eliminated a major run of landlocked Atlantic salmon in the Clyde River, which flows into Lake Memphremagog in the Northeastern Highlands. Dams all along the southern reach of the Connecticut River have similarly affected the runs of salmon and shad. These species were eliminated from the river in the nineteenth century but have since been restored, although in much-reduced numbers, by fish and game agencies.

Rivers and streams throughout the state have also been seriously altered by the dumping, accidental or intentional, of many different pollutants. Levels of septic sewage, agricultural fertilizers, road salt, gasoline, heavy metals, pesticides, and sediments of many different types have measurably increased in more than one thousand miles of these waterways in recent years, much to the detriment of the plants and animals that live there.

Ponds and Lakes

Not counting Lake Champlain, there are more than 230 lakes and ponds in Vermont, covering more than 52,000 acres. They are as diverse as the free-flowing waterways but can be divided into four basic types based on the chemistry of their waters.

Dystrophic ponds are those that have high concentrations of tannins, leached into the water from surrounding vegetation. They tend to be shallow and to occur above 1,500 feet in the southern Green Mountains, such as Grout Pond in Stratton, although they can be found at all elevations and all biophysical regions in the state. Because of the tannins, penetration of light into the water is low, and vegetation is usually at or close to the surface. Various species of pondweed, bulhead lilies, small floating manna grass, and quillwort are all typical plants in dystrophic ponds. Thick layers of undecomposed organic matter tend to develop on the bottoms of such ponds. Dystrophic ponds do not usually support a rich diversity of animals. Brown bullhead is the most common fish, although both golden shiner and brook trout can live there as well. They are successful in such ponds largely because they easily tolerate the harsh winter conditions at high elevations in Vermont.

Also generally found at high elevations are ponds formed over noncalcareous bedrock, which then cannot easily neutralize acids that enter the water from precipitation or from the surrounding soil. The water in these ponds is highly acidic, which makes it difficult for microscopic plants and animals to grow; therefore, acidic ponds are extremely clear. They accumulate little organic matter on their bottoms, although several species of aquatic plants that are tolerant of acidity grow there in small amounts, including duckgrass and water lobelia. They share many animals with dystrophic ponds. An example of such a high-elevation acidic pond is Forester Pond in Jamaica in the southern Green Mountains.

Oligotrophic lakes, so named because their waters tend to be low in nutrients available for aquatic animals, are located primarily in the Northeastern Highlands. Typical oligotrophic lakes are Seymour Lake in Morgan, Lake Willoughby in Westmore, and Little Averill Pond in Averill. They tend to be deep, often more than one hundred feet in places. The acidity of these lakes is less than that of high-elevation acidic ponds but can still be quite high, depending on the ability of the underlying bedrock to neutralize acids in the water. As a result, acidic ponds and oligotrophic lakes share many plant species, although plants occur at much greater densities in oligotrophic lakes. Because of their greater depth and colder waters, oligotrophic lakes support many species of fish that prefer cold waters, including lake trout, rainbow smelt, burbot, and round whitefish.

Mesotrophic-eutrophic lakes have moderate to high levels of nutrients (see fig. 9.5). Unlike other types of lakes and ponds, the littoral, or nearshore, portion of a mesotrophic-eutrophic lake often supports a diverse community of aquatic plants. In addition to many species of pondweed that specialize on waters with high nutrient levels, these lakes contain hornwort, duckweed, and water flaxseed. Mesotrophic-eutrophic lakes are found mostly in the Champlain Valley, and include Lake Iroquois in Hinesburg and Lake Dunmore in Salisbury. The warmer, more nutrient-rich waters in mesotrophic-eutrophic lakes lead to a very different assemblage of animals than are found in other lakes. Fish include chain pickerel, northern pike, emerald shiner, bluntnose minnow, white sucker, bluegill, pumpkinseed, yellow perch, and walleye.

Among the many species of birds and reptiles that feed on fish in lakes and ponds are the common loon, osprey, belted kingfisher, snapping turtle, and water snake. Lakes throughout Vermont also support a diversity of frogs along their margins, especially, green, wood, bull-, and pickerel frogs.

Lakes and ponds in Vermont have been modified so heavily over the past two hundred years that good examples of undisturbed lakes are difficult to find. Game fish have been stocked widely in ponds and lakes for over one hundred years (see chapter 5). Two thirds of all the lakes and ponds in Vermont have been modified to such an extent that they no longer support the full scope of their native flora and fauna. Human developments around

FIGURE 9.5. A mesotrophic-eutrophic lake. Photo by Stephen Trombulak.

lakes generally result in an increase in inputs of nutrients into the lakes from septic systems and surface flow. These lakes and ponds thus tend toward being eutrophic. Such transitions have serious impacts on the organisms in these waters. Because nutrient levels are often the biggest limitations to population growth of aquatic organisms, an increase in nutrient levels brings an increase in growth of microscopic plants. These algae grow profusely in eutrophic waters, especially close to the surface where light used for photosynthesis is most available. After they die, however, their bodies sink to the bottom, where other organisms begin the process of decomposition. Decomposition requires oxygen, and soon the levels of oxygen in the water become so depleted that few larger aquatic organisms can survive, leading to massive die-offs of fish and insects.

Another problem facing lakes and ponds in Vermont, especially mesotrophic-eutrophic lakes in the Champlain Valley, is the spread of exotic aquatic plants, particularly Eurasian milfoil. When Eurasian milfoil becomes established, it grows in such abundance that it covers vast areas of the bottom in shallow water, choking out the native aquatic plants, many of which are more valuable wildlife foods, and making the lake less attractive for humans. The ease with which Eurasian milfoil spreads is enhanced by its capacity to sprout new individuals from mere pieces of a stem. All it takes for Eurasian milfoil to become established in a lake is for one stem to be transferred from an invaded lake on the legs of aquatic birds, on the hull of a boat, or on the propeller of a motor.

At least four species of fish have been introduced to Vermont. Some were introduced on purpose as game fish. Brown trout, imported from Germany, were introduced to New York and Michigan in 1883, and from there widely throughout much of eastern North America, including Vermont. Rainbow trout were introduced to several places in eastern North America, beginning in New York in 1874, from their native range in western North America. Carp were introduced widely to lakes, ponds, and rivers in North America from Europe throughout the nineteenth century. Goldfish, close relatives of carp that often hybridize with them, were also released widely in North America, including Vermont; these releases were generally accidental, since goldfish were prized mostly as pets or as ornamentals for garden ponds.

Lake Champlain

Unique among lakes in Vermont is Lake Champlain (see fig. 9.6), which stretches for 125 miles and covers an area of almost 440 square miles. Even today, with an area much smaller than it was 12,000 years ago during the time of Lake Vermont, it is the one of largest deep, cold-water oligotrophic lakes in the United States, and has many characteristics that make it more like an inland sea. Hydrologically, Lake Champlain functions as several different bodies of water. South Lake is narrow and shallow for most of its thirty-two-mile length, making it more like a river than a lake. North of this is the Main Lake, which includes its deepest portion, four hundred feet at its maximum, and more than 80 percent of the lake's entire volume. To the northeast of the Main Lake lie three other shallower portions of Lake Champlain, separated from the Main Lake by a series of large islands. These are Mallets Bay, the Northeast Arm, and Missisquoi Bay (collectively named the Restrictive Arm). The Main Lake itself has three regions that differ from one another in terms of bottom topography: a narrow southern portion, a wider and deeper middle portion, and a narrow and shallower northern portion near where the lake drains into the Richelieu River.

The geological characteristics of Lake Champlain all represent a continuing response of the region to Vermont's geological past. The contours of the lake's bottom were carved by the lobe of the glacier that flowed through the Champlain Valley. The valley itself is a consequence of the continental collision 450 million years ago that fractured the crust and allowed the floor of the valley to drop (see chapter 1), and the lake's current water level is a response to the amount that the earth's crust has risen in this region since the glacier retreated and its massive weight was removed.

FIGURE 9.6. Lake Champlain from Chimney Point. Photo by Stephen Trombulak.

The biological characteristics of Lake Champlain also represent a response to its geological past. Eighty of the more than ninety species of fish found in Vermont occur in Lake Champlain; many of them—such as the lake sturgeon, longnose gar, bowfin, mooneye, cisco, quillback, sauger, and freshwater drum—occur nowhere else in the state. Many of the lake's fish species, such as the burbot, are characteristic of more northern climates and are found in the lake because of its deep waters and its northward connection to the Saint Lawrence River. The earlier existence of the Champlain Sea resulted in the presence of a number of species of fish that normally spend part of their lives in the open ocean, including the rainbow smelt, sea lamprey, and, historically, Atlantic salmon. Some fish species in Lake Champlain, including black bass, catfish, American eels, yellow perch, northern pike, walleye, pickerel, rock bass, smelt, Atlantic salmon, whitefish, and sturgeon, have at times been so abundant that over the last two hundred years they have supported major commercial fisheries.

Lake Champlain also supports a rich diversity of microscopic plants and animals, shrimp, insects, mussels, aquatic worms, bryozoans, and sponges. Several hundred species of invertebrates have been found in the lake, most of which specialize in one of the many different habitats the lake offers. The lake is also home to two species of turtle that are found nowhere else in the state. The common map turtle is restricted to Lake Champlain and the mouths of its major tributary rivers. The spiny softshell turtle is known in this region only from the Northeast Arm and Missisquoi Bay. Indeed, this population, along with a few others to the north in Quebec, are isolated from the species's main distribution in the Midwest, probably as a result of the most recent ice age.

The natural history of Lake Champlain has been altered in recent times by several different factors. First, studies of the lake's water indicate that no part of the lake is unaffected by pollution. Dumping from industries and towns, deposition of heavy metals and fertilizers carried into the lake by its tributaries, and atmospheric deposition of pollutants transported into the region from thousands of miles away have all taken their toll.

Second, native species have also declined from overharvesting. For example, Atlantic salmon were present in sizable numbers in Lake Champlain up until the mid-nineteenth century. They were probably extirpated as a result of intense fishing and the damming of the lake's tributary rivers. Salmon are migratory, moving from large bodies of water, such as lakes or the ocean, up into rivers to lay their eggs; without access to the upper reaches of breeding rivers flowing into Lake Champlain, the population could not replace itself and disappeared. Other species of fish have shown declines recently, especially the lake sturgeon, which might be disappearing from Lake Champlain for the same reasons as the Atlantic salmon.

Third, exotic species have disrupted the ecology of Lake Champlain. The most worrisome of these now is the zebra mussel, a tiny bivalve native to eastern Europe and accidentally transported to the Great Lakes in the 1980s, where it had established a population by 1988. Zebra mussel, only the size of a thumbnail, is a prolific breeder. Colonies spread over all kinds of substrates, including the shells of the larger native species of mussels, making it impossible for natives to open their shells and feed. Zebra mussels also filter out of the water much of the microscopic food that other aquatic invertebrates need to survive. Despite efforts to keep zebra mussels out of Lake Champlain, a colony was found in the southern portion of the lake in 1993. Since then they have spread to both the northern and southern ends of Lake Champlain. Another invader is the water chestnut, which entered southern Lake Champlain from the Hudson River in the 1970s. This plant grows in thick mats in shallow water throughout the southern half of the lake, making boating and swimming difficult. At the present, there is little hope that either zebra mussel or water chestnut can be removed from Lake Champlain, and there is no clear picture yet what the full extent of their effects on the natural history of the lake will be.

Terrestrial, wetland, and aquatic communities together create a rich biological tapestry within which the history of human events in Vermont is woven. As with terrestrial communities, wetland and aquatic communities are diverse and have been transformed over time by several cultural forces: pollution, damming, draining, overfishing, and introduction of exotics. These communities, however, have not shown the same trends toward restoration of natural abundances and distribution as have terrestrial, particularly forested, communities. Thus, the characterization of the Vermont landscape as one that has recovered over the past hundred years from earlier degradation is true in only a limited sense. Achieving the same kind of ecological restoration for wetland and aquatic communities as was begun for forested communities more than one hundred years ago is now one of our society's great challenges and will in part determine the character of Vermont's future.

PART IV

Conclusion

10

The Futures of Vermont

IT IS CLEAR that the Vermont landscape has been shaped by many forces, both natural and cultural, and that these forces have exerted their influences over different spans of time. Three broad themes emerge from this history.

To begin with, the natural history of a region, although primarily shaped by geological and ecological forces, must also include a clear recognition of human effects. During Vermont's last two hundred years, roughly three quarters of the forestland in the state was cleared, but a significant amount of these open fields has since returned to forestland. The transition in the dominant natural-community types directly or indirectly resulted in the loss of many native species (such as wolves and mountain lions), the migration into Vermont of species native elsewhere in North America but not previously found here (such as coyotes and turkey vultures), and the importation by people of species entirely exotic to the region. Similarly, the cultural history of a region must also include an appreciation of geological and biological conditions. The history of agriculture in Vermont is influenced fundamentally by the state's topography, climate, and soils, features that can occasionally be influenced by human action but are never completely under human control. Furthermore, because ecological forces led to almost all of Vermont being forested, the European Americans who came to settle there had to invest tremendous labor in clearing the forest for cropland and pasture.

The second theme is that the history of this landscape can be understood only by considering events that take place within a much broader context. Wool tariffs in Washington, D.C., air pollution from power plants in the midwestern United States, and the development of industrial-scale dairy farms in the north-central United States are but three examples of events far outside Vermont's political borders that affect its natural communities, as well as economic realities for the people living there. In the context of both natural and cultural history, borders are fuzzy, and they define only a limited part of our relationship with a landscape.

Finally, natural communities are understood only incompletely as a list of species. Species respond to and participate in ecological processes and inhabit a region as a result of long-term patterns or changes in climate and geography. The natural history of a region is the result of a complex set of geological, biological, and cultural forces that operate on entire communities, as was the case for the clearing of forestland and the draining of wetlands. Natural communities create the biological landscape within which species live. The responses of communities to ongoing change and their resilience in the face of many kinds of disturbance provide essential ingredients of the region's overall character.

Vermont's future will be shaped by geological, biological, and cultural forces, but the power to predict the future is linked directly to the time span over which one is trying to view. The major geological events that shaped the history of Vermont took place tens of thousands to millions of years ago. Vermont in the future, as shaped by future geological forces, might well differ greatly from Vermont of today. Ten thousand years from now, Vermont might once again be covered by a continental glacier a mile or more thick. One hundred million years from now, Vermont might be in the midst of another multicontinental collision or be part of a shallow sea. Over those time scales, in the face of such dramatic change, few predictions have any real relevance.

The biological forces that led to the Vermont of today, by contrast, operated on the order of thousands of years. A few thousand years from now, the natural communities in this region might be entirely different, probably representing responses to changes both in climate and in the conditions created by the human cultures that live here over the intervening years. Yet it can reasonably be predicted what natural communities will be like in this region over the next several hundred years if they are allowed to develop in their own ways and in their own times.

Cultural forces operate over the shortest time scale, a matter of tens to hundreds of years. It is within this time frame that predictions make the most sense. We see many emerging or continuing trends having powerful influences on the future history of Vermont's landscape.

First is that knowledge of the natural world, especially of the importance of biological diversity and ecological processes, has increased tremendously over the past few decades. Aldo Leopold, in his essay "The Round River," wrote, "The last word in ignorance is the man who says of an animal or plant: 'What good is it?' If the land mechanism as a whole is good, then every part is good, whether we understand it or not. If the biota, in the course of aeons, has built something we like but do not understand, then who but a fool would discard seemingly useless parts? To keep every cog and wheel is the first precaution of intelligent tinkering." From this philosophy has emerged the discipline of conservation biology, which seeks—and will continue to seek—to wed intelligently the fields of ecology, wildlife biology, and natural-resource management in order to provide guideposts for how humans can live in harmony with nature.

Second, as demonstrated by the birth and growth of the conservation movement, people increasingly realize the importance of natural communities—both for humanity's own well-being and for the well-being and inherent worth of other species. Federal and state laws, land trusts, public lands, and changing public opinion all show positive trends toward sustainable human and natural communities.

A third trend, growing out of the second, is the increasing cooperation on natural-resource and conservation issues along ecological rather than political lines. Examples of such initiatives—most of which include the federal government—include the Champlain-Adirondack Biosphere Reserve, the Lake Champlain Basin Program, the series of Northern Forest programs, and the Silvio Conte National Wildlife Refuge on the Connecticut River along its entire four-state length.

That the landscape will continue to be influenced by economic forces and conditions that originate beyond Vermont is a fourth trend. The development of NAFTA and the General Agreements on Tariffs and Trade (GATT) have serious implications—some predictable, some not—for the economics of resource production in Vermont. Multinational agribusiness and timber corporations move control of the Vermont economy outside of its borders, furthering a trend begun more than three hundred years ago with the beginning in trade of beaver pelts. Such connections with regional, national, and international economies form the foundation for one of the great ironies of Vermont's history: The return of Vermont's forests and the rise of its reputation as an environmental success story has, in fact, been made possible only by the globalization of Vermont's economy. In the middle of the nineteenth century, when Vermont was at the nadir of its forest cover, the land was supplying virtually all the food and energy for the people who lived there. The return of the forests could occur only when Vermonters were able to acquire the resources they needed by taking them from other landscapes. Today, the forest is largely back, albeit in greatly modified form, but Vermont's human population, which is nearly double that of 1850, imports almost all of its food and energy. This is a lesson that should be kept in mind as society increasingly discusses sustainable development. The globalization of the economy will also continue to bring with it a host of changes, including the continued spread of exotic plants, animals, and diseases. The future will surely include new and more versions of the chestnut blight, gypsy moth, Dutch elm disease, Eurasian milfoil, and zebra mussel.

Fifth, growth in both human-population size and resource consumption are projected to continue for the foreseeable future. More people with more expectations for access to goods and services will bring with them increased levels of air and water pollution, decreased levels of water availability, and transformations in land use through development. Vermont's population in 1995 was 585,000, giving it a population density of sixty-three people per square mile. If Vermont had a population density the same as Connecticut —something that is well within possibility given trends in technology that allow for faster communication and transportation—its population would be 6.2 million, more than ten times that of today.

A final and related trend is that technological developments will continue to influence the connections of the people in Vermont with other regions and will shape how people interact with the landscape. The Internet and fax machines have brought about an increase in telecommuting, biotechnology has changed the dairy industry, and whole-tree harvesters and aerial herbicide spraying could undo the partial recovery the state's forests have enjoyed during the last hundred years.

Many different possible futures lie before us as a society, depending on the choices we make. We began this book on top of Mount Abraham in the Green Mountains, looking out over a landscape of forested mountains and valleys, all showing signs of human occupancy. What might we see if we return to this mountaintop forty years from now? We foresee three general scenarios.

One possible future is for trends over the next forty years to not change very much, leading to a future very much like the present. Population could increase by a third, and with it would come a gradual spread of human development—houses, roads, minimarts—and a parallel decline in farmland. Forestland, both public and private, would continue to be the dominant visual image of Vermont's landscape and, except at the highest elevations, would continue to be modified significantly by timber extraction. Exotic species would continue to spread, and native species would continue to respond individually and as communities to the host of forces that determine their ability to survive in an evolving landscape. The ecological recovery of the Vermont landscape—the return of older forests, moose, and healthy bear populations—would be slowly reversed under this scenario, as natural communities would be altered even more to meet the needs of human communities.

A second possible future is one of hyperdevelopment. The pastoral and forested landscape of Vermont today could become transformed into the kind of developed landscape that is currently characteristic of southern New England. In this scenario, Burlington would develop into a major urban center, and towns such as Rutland and Brattleboro would become midsized cities. Major transportation corridors, such as Interstate 89 and Route 7, would pass through miles of strip development, and new major corridors would be developed, especially from east to west. Private forestland and farmland would succumb to development, and the only places where the original natural communities of Vermont would have a chance to persist would be in the Green Mountain National Forest, the few state parks, and the Nature Conservancy's preserves.

A third possible future involves the creation of dominant-use zones throughout the state, located and apportioned so that Vermont in 2040 would contain all of the elements that in the 1990s are considered vital for both natural and cultural communities. The landscape would reflect a matrix of land-use categories in which different uses would predominate: urban centers, working farms and forests, and wild natural communities. Each land-use category would be located and connected to maximally achieve its best and highest goals, and each apportioned a fair share of the landscape so that the needs of all Vermonters and all native species could be met.

This scenario would allow many of the elements of the Vermont of four hundred years ago to return. Areas designated for wild nature would once again exist as a shifting mosaic of natural-community types and ages, where a preponderance of the forests would be in old-growth condition. Immense trees in cathedral-like stands would provide habitat for a startling array of native plants and animals, some of which, like the wolf, mountain lion, and American chestnut, have been absent for more than one hundred years. Exotic species would still remain, being impossible to completely eradicate, but their numbers would be reduced greatly by the decline in open fields. Within these reserves, roads would be obliterated and waters would run free, clean, and rich with native aquatic plants and animals. These reserves would be part of a larger continentwide system of wild lands and would contribute to the restoration of ecological health far outside of Vermont's political borders.

The working landscape would continue to function as it has for the past 250 years, but now trees and crops would be harvested increasingly as parts of locally based, value-added economic networks. Agricultural and forestry practices would promote ecological sustainability, increasing opportunities for future generations to live and work on the land as people have done here in one way or another for 10,000 years. Family forests and farms would again be the dominant model of land stewardship, providing the nuclei for rural communities throughout the state. Urban centers such as Burlington would continue to exist as areas of high human-population density and would increasingly serve as the state's growth centers for light industry and human populations to allow rural and wild areas to persist. To the greatest extent possible, the economy would be diversified and kept local to minimize the kinds of disruptive external forces that have buffeted Vermont since the beginning of European colonization.

The future is only incompletely under our control. Humanity has no choice but to be a part of the larger web of connections that link us as a species to the rest of the natural world, and local culture has no choice but to be a part of the larger human society. Events that take place far outside of Vermont, or even the Greater Laurentian Region, will help set the stage on which the future will be played. But neither is the future completely out of our control. We can choose how we wish to deal with questions about urban development and natural-resource management. We can decide to address each question before us one at a time, each one isolated from the larger issue of what we want Vermont to look like in 2040. Or we can decide to engage in and defend vigorously efforts at long-range planning for our communities and for the conservation of wild nature. We can engage in debates about our future as a society divided into competing special-interest groups, aggressively defending what we perceive to be our needs against all others. Or we can engage in debates as neighbors, working together toward a future that we will be proud to pass on.

We have no crystal ball to show us what the future *will* look like, but we feel confident that society's ability to achieve the future that it wants—one that includes both healthy natural and healthy human communities, depends on all its members remaining aware of our connections to each other as people, our connections as a species to the natural world, our capacities to bring about change, and our capacities to respond to the world as it changes.

The forces that have shaped Vermont's past will shape its future. Vermont is a part of a vast natural and cultural history being played out in a landscape that has no distinct borders. The story of Vermont will continue to be shaped by a complicated set of natural and cultural forces continuously operating over time. History has shown that Vermont in all of its dimensions—the landscape, the natural communities and their component species, and the people—is adaptive when needs demand and resilient in the face of constant change. This is Vermont's true character.

Sources for Illustrations and Tables

Map 1.1: Adapted from information compiled by the Wisconsin Ice Age Park and Trail Foundation (http://www.iceagetrail.org/wgl.htm).

Map 2.1: Charles G. Doll, 1970, Surficial Geologic Map of Vermont, Waterbury, Vt.: Vermont Geological Survey; Glenn E. Myer and Gerhard K. Gruendling, 1979, *Limnology of Lake Champlain*, Burlington: Lake Champlain Basin Study, p. 18.; David P. Stewart and Paul MacClintock, 1969, *The Surficial Geology and Pleistocene History of Vermont*, bulletin no. 31, Montpelier: Department of Water Resources, pp. 101–11, 160–79.

Map 2.2: Myer and Gruendling, *Limnology of Lake Champlain*, p. 18; Stewart and MacClintock, *The Surficial Geology and Pleistocene History of Vermont*, pp. 179–83.

Map 2.3: Northern Cartographic, South Burlington, Vt.

Map 2.4: William A. Haviland and Marjory W. Power, 1994, *The Original Vermonters: Native Inhabitants, Past and Present*, rev. ed., Hanover, N.H.: University Press of New England, pp. 26, 46, 88, 112, 136; Dean R. Snow, 1980, *The Archeology of New England*, New York: Academic Press, pp. 104, 167, 189, 263, 318; J. V. Wright, 1979, *Quebec Prehistory*, Toronto: Van Nostrand Press, pp. 22, 26, 51, 63.

Map 2.5: Bruce G. Trigger, ed., 1978, *Handbook of North American Indians*, vol. 15: *Northeast*, Washington: Smithsonian Institution Press, p. ix.

Map 2.6: Vermont Biodiversity Project, unpubl. data; U.S. Forest Service, Forest Health Protection GIS Group, 1995, Map of the Eco-regions of New England and New York, Durham, N.H.

Map 2.7: U.S. Geological Survey, 1974, Hydrologic Unit Map—1974: States of New Hampshire and Vermont, Reston, Va.: U.S. Geological Survey.

Map 3.1: Kenneth T. Jackson and James T. Adams, eds., 1978, *Atlas of American History*, rev. ed., New York: Scribner's, p. 76.

Map 3.2: Marcus L. Hansen, 1940, *The Mingling of the Canadian and American Peoples*, vol. 1, New Haven, Conn.: Yale University Press, p. 90.

Map 3.3: Edward Fuller, 1952, *Vermont: A History of the Green Mountain State*, Montpelier: State Board of Education, p. 116; Peter S. Onuf, 1983, *The Origins of the Federal Republic: Jurisdictional Controversies in the United States, 1775–1787*, Philadelphia: University of Pennsylvania Press, p. 126.

Map 4.1: Thomas Arnold, 1981, "Two Hundred Years and Counting: Vermont

Community Census Totals, 1791 to 1980," Burlington: University of Vermont Center for Rural Studies, pp. 15–32; Maurice Saint-Yves, 1982, *Atlas de Géographie Historique du Canada*, Boucherville, Quebec: Les Éditions Françaises, p. 43 (data for 1861); U.S. Bureau of the Census, 1853, *The Seventh Census of the United States: 1850*, Washington: Robert Armstrong, pp. 20–21, 30–32, 93–108.

Map 4.2: T. D. Bassett, 1982, "500 Miles of Trouble and Excitement: Vermont's Railroads, 1848–1861," in *In a State of Nature: Readings in Vermont History*, ed. H. Nicholas Muller and Samuel B. Hand, Montpelier: Vermont Historical Society, p. 164; Harold F. Wilson, 1967 [1936], *The Hill Country of Northern New England: Its Social and Economic History, 1790–1930*, New York: AMS Press, p. 37.

Map 5.1: *Canada Gazatteer Atlas*, 1980, Macmillan; Fuller, *Vermont*, p. 190; National Inventory of Dams—Vermont (Nick Forbes, U.S. Army Corps of Engineers, New England Division); *The New Hampshire Atlas and Gazetteer*, 1988, Freeport, Me.: DeLorme Mapping; *The New York Atlas and Gazetteer*, 1988, Freeport, Me.: DeLorme Mapping; *Rand McNally Road Atlas*, 1997, Chicago: Rand McNally.

Map 6.1: Statistics Canada, 1997, *GeoRef: 1996 Census of Canada*, Ottawa: Industry Canada, CD-ROM (Canadian data for 1996); U.S. Department of Commerce, Bureau of Census, 1992, *1990 Census of Population*, Washington: Government Printing Office, CD-ROM.

Map 6.2: *Rand McNally Road Atlas*, 1997, Chicago: Rand McNally, p. 98; Vermont Chamber of Commerce, 1998, *Vermont Winter Guide*, Montpelier: Vermont Chamber of Commerce, pp. 39, 70, 121.

Map 6.3: New York State Commission on the Adirondacks in the Twenty-first Century, 1990, Adirondack Park Open Space Protection Plan Map, Albany: New York State Commission on the Adirondacks in the Twenty-first Century; Society for the Protection of New Hampshire Forests, 1994, Conservation/Public Lands GIS Coverage, New Hampshire GRANIT Net (http://nhresnet.sr.edu/granit/data-overview.html); Vermont Agency of Natural Resources, 1993, Publand GIS Coverage, Vermont Center for Geographic Information (http://geo-vt.uvm.edu/).

Table 4.1: Muller and Hand, *In a State of Nature*, p. 401 (human population); Harold A. Meeks, 1986, *Time and Change in Vermont: A Human Geography*, Chester, Conn.: Globe Pequot Press, p. 159 (sheep and cow population); Lewis D. Stilwell, 1948, *Migration from Vermont*, Montpelier: Vermont Historical Society, p. 158 (1828 sheep population).

Table 5.1: Muller and Hand, *In a State of Nature*, pp. 401–2 (human population); Meeks, *Time and Change in Vermont*, p. 159 (sheep and cow population); U.S. Department of Agriculture, 1951, *Agricultural Statistics 1951*, Washington: Government Printing Office, p. 354 (1950 sheep population).

Table 5.2: *Farmland*: U.S. Department of the Interior, *Census Office, Eleventh Census of the United States: Report on the Statistics of Agriculture in the United States* (1895), p. 232 (1890); U.S. Census Office, *Twelfth Census of the United States: Agriculture:* part 1 (1902), p. 142 (1900); *Fourteenth Census of the United States: Agriculture:* vol. 5 (1922), p. 36 (1880, 1910, 1920); *Sixteenth Census of the United States: Agriculture:* vol. 3 (1943), p. 38 (1930, 1940); *U.S. Census of Agriculture: General Report:* vol. 2 (1950), p. 32 (1950), Washington: Government Printing Office. *Forestland*: U.S. Department of the Interior, Census Office, *Tenth Census of the United States: Report on the Forests of North America:* vol. 9 (1884), pp. 498–500 (1880); U.S. Census Office, *Twelfth Census of the United States: Manufactures:* part 3 (1902), p. 834 (1900), Washington: Government Printing Office; Wesley Bradfield, 1909, "Standing Timber in Wood Lots," in *Report of the National Conservation Commission:* vol. 2, U.S. Senate,

60th Congress, 2d Session, p. 185 (1909); Ralph Widner, ed., 1968, *Forests and Forestry in the American States*, Missoula, Mont.: National Association of State Foresters, p. 570 (1930); Tom Frieswick, USDA Forest Service, Northeastern Area State and Private Forestry, Radnor, Pa. (personal communication, 1998) (1938); John R. McGuire and Robert D. Wray, 1952, *Forest Statistics for Vermont*, Upper Darby, Pa.: USDA, Forest Service Northeast Forest Experiment Station, p. 1 (1948); Lloyd Irland, 1982, *Wildlands and Woodlots: The Story of New England's Forests*, Hanover, N.H.: University Press of New England, p. 115 (1952).

Table 6.1: Muller and Hand, *In a State of Nature*, p. 402 (human population through 1980); U.S. Department of Commerce, Bureau of Census, 1992, *1990 Census of Population: General Population Characteristics: Vermont*, Washington: Government Printing Office, p. 1 (human population 1990); U.S. Department of Commerce, Bureau of Census, 1996, *Statistical Abstract of the United States: 1996*, Washington: Government Printing Office, p. 28 (human population 1995); Meeks, *Time and Change in Vermont*, p. 159 (cow [1960 and 1970] and sheep populations [1970]); United States Department of Agriculture, *Agricultural Statistics 1961* (1962), p. 338, *Agricultural Statistics 1981* (1981), pp. 304, 325, *Agricultural Statistics 1991* (1991), pp. 248, 271, *Agricultural Statistics 1995–1996* (1995–1996), pp. VII-3, VII-31, Washington: Government Printing Office (cow [1980 through 1995] and sheep populations [1960 and 1980 through 1995]).

Table 6.2: *Farmland*: United States Department of Commerce, Bureau of the Census, 1962, *U.S. Census of Agriculture 1959:* vol. 1: part 3, *Counties: Vermont*, Washington: Government Printing Office, p. 3; United States Department of Agriculture, *Agricultural Statistics 1971* (1971), p. 441, *Agricultural Statistics 1981* (1981), p. 416, *Agricultural Statistics 1991* (1991), p. 355, *Agricultural Statistics 1995–1996* (1995–1996), p. IX-4, Washington: Government Printing Office. *Forestland*: Tom Frieswick, USDA Forest Service, Northeastern Area State and Private Forestry, Radnor, Pa. (personal communication, 1998).

Table 6.3: Preston Bristow, Vermont Land Trust (personal communication, 1998), and Ed Leary, Lands Administrator, Department of Forests, Parks and Recreation, Vermont Agency for Natural Resources (personal communication, 1998).

Table 7.1: Modified from Elizabeth Thompson, 1997, *Natural Communities of Vermont*, private printing for the Vermont Nature Conservancy.

Table 8.1: Modified from Thompson, *Natural Communities of Vermont*.

Table 9.1: Modified from Thompson, *Natural Communities of Vermont*; Aquatics Classification Workgroup, 1998, *A Classification of the Aquatic Communities of Vermont*, private printing prepared for the Vermont Nature Conservancy and the Vermont Biodiversity Project.

Figure 1.1: Frank Press and Raymond Siever, 1978, *Earth*, San Francisco: W. H. Freeman Press, pp. 457–81.

Figure 1.2: Y. W. Isachsen, E. Landing, J. M. Lauber, L. V. Rickard, and W. B. Rogers, eds., 1991, *Geology of New York: A Simplified Account*, Educational Leaflet no. 28, Albany: New York State Museum/Geological Survey, The State Education Department, modified from figs. A.4-2, A.4-5, A.4-29, A.4-36, A.4-40, A.4-41, A.4-45, A.4–53, and A.4–60, created by Barbara Tewksbury.

Figure 4.1: Roland M. Harper, 1918, "Changes in the Forest Area of New England in Three Centuries," *Journal of Forestry*, 16:447; data from tables 5.2 and 6.2.

References

Geology of Vermont

Isachsen, Y. W., E. Landing, J. M. Lauber, L. V. Rickard, and W. B. Rogers, eds. 1991. *Geology of New York: A Simplified Account*. Educational Leaflet no. 28. Geological Survey, the State Education Department. Albany: New York State Museum.

Myer, Glenn E., and Gerhard K. Gruendling. 1979. *Limnology of Lake Champlain*. Burlington: Lake Champlain Basin Study.

Press, Frank, and Raymond Siever. 1978. *Earth*. San Francisco: W. H. Freeman.

Raymo, Chet, and Maureen E. Raymo. 1989. *Written in Stone: A Geological History of the Northeastern United States*. Chester, Conn.: Globe Pequot.

Redfern, Ron. 1983. *The Making of a Continent*. New York: Times Books.

Stewart, David P., and Paul MacClintock. 1969. *The Surficial Geology and Pleistocene History of Vermont*. Bulletin no. 31. Montpelier: Department of Water Resources.

Van Diver, Bradford B. 1987. *Roadside Geology of Vermont and New Hampshire*. Missoula: Mountain Press.

Welby, Charles W. 1962. *Paleontology of the Champlain Basin in Vermont*. Special publication no. 1. Montpelier: Vermont Geological Survey.

Biology of Vermont

Aquatics Classification Workgroup. 1998. *A Classification of the Aquatic Communities of Vermont*. Prepared for the Vermont Nature Conservancy and the Vermont Biodiversity Project. Private printing.

Bell, Nancy. 1990. "Vermont's Elusive Bears." *Vermont Environmental Report* (Spring/summer): 12–14.

Borror, Donald J., and Richard E. White. 1970. *A Field Guide to the Insects of America North of Mexico*. Boston: Houghton Mifflin.

Conant, Roger, and Joseph T. Collins. 1991. *Reptiles and Amphibians: Eastern/Central North America*. Boston: Houghton Mifflin.

Davis, Mary Byrd. 1993. *Old Growth in the East: A Survey*. Richmond, Vt.: Cenozoic Society.

Davis, R. B., and G. L. Jacobson Jr. 1985. "Late Glacial and Early Holocene Landscapes in Northern New England and Adjacent Areas of Canada." *Quaternery Research* 23:341–68.

DeGraaf, Richard M., and Deborah D. Rudis. 1983. *Amphibians and Reptiles of New England: Habitats and Natural History*. Amherst, Mass.: University of Massachusetts Press.

DeGraff, Richard M., Mariko Yamasaki, William B. Leak, and John W. Lanier. 1992. *New England Wildlife: Management of Forested Habitats*. U.S. Department of Agriculture, Forest Service, Northeastern Forest Experimental Station, General Technical Report NE-144. Washington: Government Printing Office.

Fichtel, Christopher, and Douglas G. Smith. 1995. *The Freshwater Mussels of Vermont*. Nongame and Natural Heritage Program, Vermont Fish and Wildlife Department Technical Report no. 18. Montpelier: Leahy Press.

Foster, D. R., J. D. Aber, J. M. Melillo, R. D. Bowden, and F. A. Bazzaz. 1997. "Forest Response to Disturbance and Anthropogenic Stress." *BioScience* 47:437–45.

Godin, Alfred J. 1977. *Wild Mammals of New England*. Chester, Conn.: DeLorme Publishing.

Johnson, Charles W. 1985. *Bogs of the Northeast*. Hanover, N.H.: University Press of New England.

———. 1980. *The Nature of Vermont*. Hanover, N.H.: University Press of New England.

Laughlin, Sarah B., and Douglas P. Kibbe, eds. 1985. *The Atlas of Breeding Birds of Vermont*. Hanover, N.H.: University Press of New England.

Linder, Will. 1994. "The Clyde—A River of the Kingdom." *Vermont Environmental Report* (Fall): 19–22.

Loucks, Orie L. 1998. "The Epidemiology of Forest Decline in Eastern Deciduous Forests." *Northeastern Naturalist* 5:143–54.

Marchand, Peter J. 1987. *Life in the Cold: An Introduction to Winter Ecology*. Hanover, N.H.: University Press of New England.

———. 1987. *North Woods: An Inside Look at the Nature of Forests in the Northeast*. Boston: Appalachian Mountain Club.

Mech, L. David. 1970. *The Wolf: The Ecology and Behavior of an Endangered Species*. Minneapolis: University of Minnesota Press.

Myer, Glenn E., and Gerhard K. Gruendling. 1979. *Limnology of Lake Champlain*. Burlington: Lake Champlain Basin Study.

National Geographic Society. 1983. *Field Guide to the Birds of North America*. Washington: National Geographic Society.

Perry, David A. 1994. *Forest Ecosystems*. Baltimore: The Johns Hopkins University Press.

Pielou, E. C. 1991. *After the Ice Age: The Return of Life to Glaciated North America*. Chicago: University of Chicago Press.

Popp, Robert, and Everett Marshall. 1996. "Vermont's Rare and Uncommon Native Plants." Report for the Nongame and Natural Heritage Program, Vermont Department of Fish and Wildlife. Waterbury, Vt.: Agency of Natural Resources.

Scott, W. B., and E. J. Crossman. 1973. *Freshwater Fishes of Canada*. Bulletin no. 184. Ottawa: Fisheries Research Board of Canada.

Seymour, Frank C. 1982. *The Flora of New England: A Manual for the Identification of All Vascular Plants including Ferns and Fern Allies Growing without Cultivation in New England*. 2d ed., 4th printing with supplement. Phytologia Memoirs, vol. 5.

Stern, Kingsley R. 1991. *Introductory Plant Biology*. 5th ed. Dubuque, Iowa: William C. Brown.
Teillon, H. Brenton, Barbara S. Burns, and Ronald S. Kelley. 1997. *Forest Insect and Disease Conditions in Vermont: 1996*. Waterbury, Vt.: Agency of Natural Resources.
Thompson, Elizabeth. 1997. *Natural Communities of Vermont*. Prepared for the Vermont Nature Conservancy. Private printing.
Thompson, Zadock. 1972 [1853]. *Natural History of Vermont*. Rutland, Vt.: Charles E. Tuttle.
U.S. Congress, Office of Technology Assessment. 1993. *Harmful Non-Indigenous Species in the United States*. OTA-F-565. Washington: Government Printing Office.
Vermont Reptile and Amphibian Scientific Advisory Group. 1995. *A Preliminary Atlas of the Reptiles and Amphibians of Vermont*. Middlebury, Vt.: Private printing.
Weaver, J. L., P. C. Paquet, and L. F. Ruggiero. 1996. "Resilience and Conservation of Large Carnivores in the Rocky Mountains." *Conservation Biology* 10:964–76.
Werner, Robert G. 1980. *Freshwater Fishes of New York State: A Field Guide*. Syracuse, N.Y.: Syracuse University Press.
Wessels, Tom. 1997. *Reading the Forested Landscape: A Natural History of New England*. Woodstock, Vt.: Countryman Press.
Whitney, Gordon G. 1994. *From Coastal Wilderness to Fruited Plain: A History of Environmental Change in Temperate North America from 1500 to the Present*. New York: Cambridge University Press.

Humans and the Landscape in Vermont

Aiken, George D. 1938. *Speaking from Vermont*. New York: Stokes.
Anderson, S. Axel, and Florence M. Woodard. 1932. "Agricultural Vermont." *Economic Geography* 8:12–42.
Barron, Hal S. 1984. *Those Who Stayed Behind: Rural Society in Nineteenth-Century New England*. New York: Cambridge University Press.
Bassett, T. D. Seymour. 1992. *The Growing Edge: Vermont Villages, 1840–1880*. Montpelier: Vermont Historical Society.
Bellesiles, Michael A. 1993. *Revolutionary Outlaws: Ethan Allen and the Struggle for Independence on the Early American Frontier*. Charlottesville: University Press of Virginia.
Bidwell, Percy W. 1916. "Rural Economy in New England at the Beginning of the Nineteenth Century." *Transactions of the Connecticut Academy of Arts and Sciences* 20:241–399.
Bidwell, Percy W., and John I. Falconer. 1925. *History of Agriculture in the Northern United States, 1620–1860*. Washington: Carnegie Institution.
Black, John D. 1950. *The Rural Economy of New England*. Cambridge, Mass.: Harvard University Press.
Browne, Charles C., and Howard B. Reed, eds. 1985. *Visions, Toil and Promise: Man in Vermont's Forests*. Saint Johnsbury, Vt.: Fairbanks Museum and Planetarium.
Bryan, Frank. 1974. *Yankee Politics in Rural Vermont*. Hanover, N.H.: University Press of New England.
Bryan, Frank, and John McClaughry. 1989. *The Vermont Papers: Recreating Democracy on a Human Scale*. Chelsea, Vt.: Chelsea Green.
Cronon, William. 1983. *Changes in the Land: Indians, Colonists, and the Ecology of New England*. New York: Hill and Wang.

DeGrove, John M., with the assistance of Deborah A. Miness. 1992. *The New Frontier for Land Policy: Planning and Growth Management in the States*. Cambridge, Mass.: Lincoln Institute of Land Policy.

Douglass, John A. 1981. "The Green Mountains of Vermont and Its National Forest." M.A. thesis, University of California, Santa Barbara.

Doyle, William. 1994. *The Vermont Political Tradition: And Those Who Helped Make It*. Rev. ed. Barre, Vt.: William Doyle.

Elder, John. 1998. *Reading the Mountains of Home*. Cambridge, Mass.: Harvard University Press.

Foote, Leonard E. 1946. *A History of Wild Game in Vermont*, 3d ed. Montpelier: Vermont Fish and Game Service.

Fuller, Edmund. 1952. *Vermont: A History of the Green Mountain State*. Montpelier: Vermont State Board of Education.

Haviland, William A., and Marjory W. Power. 1994. *The Original Vermonters: Native Inhabitants, Past and Present*. Rev. ed. Hanover, N.H.: University Press of New England.

Healy, Robert G., and John S. Rosenberg. 1979. *Land Use and the States*. 2d ed. Baltimore: The Johns Hopkins University Press for Resources for the Future.

Huffman, Benjamin L. 1974. *Getting Around Vermont*. Burlington: University of Vermont Environmental Program.

Irland, Lloyd C. 1982. *Wildlands and Woodlots: The Story of New England's Forests*. Hanover, N.H.: University Press of New England.

Jones, Matt B. 1939. *Vermont in the Making, 1750–1777*. Cambridge, Mass.: Harvard University Press.

Judd, Richard M. 1979. *The New Deal in Vermont: Its Impact and Aftermath*. New York: Garland.

Judd, Richard W. 1997. *Common Lands, Common People: The Origins of Conservation in Northern New England*. Cambridge, Mass.: Harvard University Press.

Klyza, Christopher McGrory. 1998. "Bioregional Possibilities in Vermont." In *Bioregionalism*, ed. Michael Vincent McGinnis. London: Routledge.

——. 1994. "Lessons from the Vermont Wilderness." *Wild Earth* (Spring): 75–79.

Klyza, Christopher McGrory, and Stephen C. Trombulak, eds. 1994. *The Future of the Northern Forest*. Hanover, N.H.: University Press of New England.

Lamson, Genieve. 1931. *A Study of Agricultural Populations in Selected Vermont Towns*. Burlington: Vermont Commission on Country Life.

Leary, Ed. 1996–1997. "Nature's Public Properties: Protecting Special Places Is a Vermont Tradition." *The Oak Log: Newsletter of the Nature Conservancy of Vermont* (Winter): 1–4.

Leopold, Aldo. 1966 [1949]. *A Sand County Almanac*. New York: Oxford University Press.

Leuchtenburg, William E. 1953. *Flood Control Politics: The Connecticut River Valley Problem, 1927–1950*. Cambridge, Mass.: Harvard University Press.

Long, John H., ed., comp. by Gordon DenBoer with George E. Goodridge. 1993. *Atlas of Historical County Boundaries: New Hampshire, Vermont*. New York: Simon and Schuster.

Marsh, George Perkins. 1965 [1864]. *Man and Nature*. Cambridge, Mass.: Harvard University Press, Belknap Press.

Mazuzan, George T. 1972. "'Skiing Is Not Merely a Schport': The Development of

Mount Mansfield as a Winter Recreation Area, 1930–1955." *Vermont History* 40:47–63.
McCullough, Robert. 1995. *The Landscape of Community: A History of Communal Forests in New England*. Hanover, N.H.: University Press of New England.
McKibben, Bill. 1995. *Hope, Human and Wild: True Stories of Living Lightly on the Earth*. Boston: Little, Brown.
Meeks, Harold A. 1986. *Time and Change in Vermont: A Human Geography*. Chester, Conn.: Globe Pequot.
———. 1986. *Vermont's Land and Resources*. Shelburne, Vt.: New England Press.
Merchant, Carolyn. 1989. *Ecological Revolutions: Nature, Gender, and Science in New England*. Chapel Hill, N.C.: University of North Carolina Press.
Merrill, Perry H. 1959. *History of Forestry in Vermont, 1909–1959*. Montpelier: State Board of Forests and Parks.
———. 1987. *Vermont Skiing: A Brief History of Downhill and Cross Country Skiing*. Montpelier: Perry Merrill.
Morrissey, Charles T. 1981. *Vermont: A History*. New York: W. W. Norton.
Muller, H. Nicholas. 1984. *From Ferment to Fatigue? 1870–1900: A New Look at the Neglected Winter of Vermont*. Occasional Paper no. 7. Burlington: University of Vermont.
Muller, H. Nicholas, and Samuel B. Hand, eds. 1982. *In a State of Nature: Readings in Vermont History*. Montpelier: Vermont Historical Society.
Newton, Earle. 1949. *The Vermont Story: A History of the People of the Green Mountain State, 1749–1949*. Montpelier: Vermont Historical Society.
Pielou, E. C. 1991. *After the Ice Age: The Return of Life to Glaciated North America*. Chicago: University of Chicago Press.
Rolando, Victor R. 1992. *Two Hundred Years of Soot and Sweat: The History and Archeology of Vermont's Iron, Charcoal, and Lime Industries*. Burlington: Vermont Archaeological Society.
Rozwenc, Edwin C. 1981. *Agricultural Policies in Vermont, 1860–1945*. Montpelier: Vermont Historical Society.
Russell, Howard S. 1976. *A Long, Deep Furrow: Three Centuries of Farming in New England*. Hanover, N.H.: University Press of New England.
Schnidman, Frank, Michael Smiley, and Eric G. Woodbury. 1990. *Retention of Land for Agriculture: Policy, Practice, and Potential in New England*. Cambridge, Mass.: Lincoln Institute of Land Policy.
Sherman, Joe. 1991. *Fast Lane on a Dirt Road: Vermont Transformed, 1945–1990*. Woodstock, Vt.: Countryman Press.
Sherman, Michael, ed. 1991. *A More Perfect Union: Vermont Becomes a State, 1777–1816*. Montpelier: Vermont Historical Society.
Silverstein, Hannah. 1995. "No Parking: Vermont Rejects the Green Mountain Parkway." *Vermont History* 63:133–57.
Steinberg, Theodore. 1991. *Nature Incorporated: Industrialization and the Waters of New England*. New York: Cambridge University Press.
Stilwell, Lewis D. 1948. *Migration from Vermont*. Montpelier: Vermont Historical Society.
Sykes, James G. 1964. "Trends in Vermont Agriculture." *Vermont Resources Research Center Report 7*. Burlington: University of Vermont Agricultural Experiment Station.
Vermont Agency of Natural Resources. 1998. *Environment 1998: An Assessment of the*

Quality of Vermont's Environment. Waterbury, Vt.: Vermont Agency of Natural Resources.

———. 1997. *Environment 1997: An Assessment of the Quality of Vermont's Environment*. Waterbury, Vt.: Vermont Agency of Natural Resources.

———. 1996. *Environment 1996: An Assessment of the Quality of Vermont's Environment*. Waterbury, Vt.: Vermont Agency of Natural Resources.

Versteeg, Jennie G. 1990. "Aspects of the Vermont-Canada Forest Products Relation in the Twentieth Century." *Vermont History* 58:164–78.

Whitney, Gordon G. 1994. *From Coastal Wilderness to Fruited Plain: A History of Environmental Change in Temperate North America from 1500 to the Present*. New York: Cambridge University Press.

Williamson, Chilton. 1949. *Vermont in Quandary: 1763–1825*. Montpelier: Vermont Historical Society.

Wilson, Harold F. 1967 [1936]. *The Hill Country of Northern New England: Its Social and Economic History, 1790–1930*. New York: AMS Press.

Woodard, Florence M. 1936. *The Town Proprietors in Vermont: The New England Town Proprietorship in Decline*. New York: Columbia University Press.

Index

Page numbers for illustrations are in *italic* type.

UNIVERSITY PRESS OF NEW ENGLAND publishes books under its own imprint and is the publisher for Brandeis University Press, Dartmouth College, Middlebury College Press, University of New Hampshire, Tufts University, and Wesleyan University Press.

Library of Congress Cataloging-in-Publication Data
Klyza, Christopher McGrory.
The story of Vermont : a natural and cultural history / Christopher McGrory Klyza and Stephen C. Trombulak.
p. cm. — (Middlebury bicentennial series in environmental studies)
Includes bibliographical references.
ISBN 0–87451–743–5 (cl : alk. paper). — ISBN 0–87451–936–5 (pbk. : alk. paper)
1. Vermont—History. 2. Natural history—Vermont. 3. Landscape—Vermont—History. 4. Vermont—Environmental conditions.
I. Trombulka, Stephen C. II. Title. III. Series.
F49.K59 1999
974.3—dc21 99–19489